Information and Meaning

An Evolutionary Perspective

AF684534

Springer
London
Berlin
Heidelberg
New York
Barcelona
Budapest
Hong Kong
Milan
Paris
Santa Clara
Singapore
Tokyo

The other books in this trilogy by Tom Stonier are:

Information and the Internal Structure of the Universe:
An Exploration into Information Physics
ISBN 3-540-19878-4

Beyond Information:
The Natural History of Intelligence
ISBN 3-540-19654-4

Tom Stonier

Information and Meaning

An Evolutionary Perspective

Springer

Professor Tom Stonier, BA, MSc, PhD, FRSA
5 The Avenue
Great Barrington
MA 01230
USA

ISBN 3-540-76139-X Springer-Verlag Berlin Heidelberg New York

British Library Cataloguing in Publication Data
Stonier, Tom, 1927-
Information and meaning : an evolutionary perspective
1.Knowledge, Theory of
I.Title
001
ISBN 354076139X

Library of Congress Cataloging-in-Publication Data
Information and meaning : an evolutionary perspective / Tom Stonier.
p. cm.
Includes bibliographical references and index.
ISBN 3-540-76139-X (pbk. : alk. paper)
1. Artificial intelligence. 2. Information theory. 3. Meaning (Philosophy) 4. Brain--Evolution. 5. Intellect. I. Title.
Q335.S88 1997 97-16365
006.3--dc21 CIP

Apart from any fair dealing for the purposes of research or private study, or criticism or review, as permitted under the Copyright, Designs and Patents Act 1988, this publication may only be reproduced, stored or transmitted, in any form or by any means, with the prior permission in writing of the publishers, or in the case of reprographic reproduction in accordance with the terms of licences issued by the Copyright Licensing Agency. Enquiries concerning reproduction outside those terms should be sent to the publishers.

© Tom Stonier 1997
Printed in Great Britain

The use of registered names, trademarks, etc. in this publication does not imply, even in the absence of a specific statement, that such names are exempt from the relevant laws and regulations and therefore free for general use.

The publisher makes no representation, express or implied, with regard to the accuracy of the information contained in this book and cannot accept any legal responsibility or liability for any errors or omissions that may be made.

Typeset by Ian Kingston Editorial Services, Nottingham, UK
Printed and bound by the Athenæum Press Ltd, Gateshead, Tyne & Wear
34/3830-543210 Printed on acid-free paper

Dedication

To Judith, who added meaning to my life

Apologia

Much of the material covered in this book draws on the work of thousands of scholars and researchers. In covering an interdisciplinary set of topics such as the evolution of information, of intelligence and of meaning, a single author, labouring in the wilds of western Massachusetts, is bound to be remiss in covering all the relevant literature.

Actually, the problem has been two-fold. On the one hand, my residence (to which I have happily retired) has no university library within an easy drive. Nor, am I able, as a matter of course, to seek out my fellow academics for advice and in-depth discussions. However, that has been the lesser of two problems. More serious has been the flood of information which has come to overwhelm me in the form of journal subscriptions, book clubs, the Internet and the occasional attendance at conferences.

The inspiration for this ***apologia*** has been the discovery (as I was meeting the deadline of sending off this manuscript to the publisher) of piles of books, journals, reprints and folders of personal notes, all stacked carefully to be used for writing this book, but left unused. It is not as though I had not read the material. In many instances, I had done so most laboriously, taking copious notes and making summaries. Unfortunately, the writing of this book was started a good many years ago, and in the meantime, under the flood of new information, much of what was valuable became misplaced and even forgotten.

I blush as I rediscover this material. I make my apologies to those scholars and researchers of whose work, in my ignorance, I was not aware. But even more so, I beg charity from my friends and colleagues of whose work I *was* aware: who sent me reprints, corresponded with me, gave advice, alerted me to other important works, and yet, somehow, fell in between the cracks of my intellectual net.

Tom Stonier
1 January 1997

Contents

· 1 ·

Introduction

Philosophical Preface

The present work is the third in a series designed to clarify the issues relating to the concept of information. The ultimate aim of these travails is to help develop a science of information. The first book, *Information and the Internal Structure of the Universe*, defined information in a manner parallel to the definitions which apply to energy: information, like energy, is conceived of as a basic property of the universe; and like energy, which is traditionally defined operationally as possessing the capacity to perform work, so information is defined operationally as possessing the capacity to organize a system.

What is being attempted in this series on information is to study a pattern of interrelated phenomena which extends through all of *nature* – patterns which range from subatomic structure to the human mind and human society – even to the abstract products of that mind and society: language and mathematics. This pattern consist of the manifestation of *order*, to a greater or lesser degree, in everything we perceive, create, or conceive. This does not preclude the possibility of systems which lack order, but order appears to be a universal phenomenon, and wherever we look for it in our universe we find it.

This question of order in the universe has preoccupied philosophers for millennia. Aristotle considered order to be part of reality: the human mind, through the senses, discovers this order. In contrast, Immanuel Kant considered that order was a product of the human mind: it is we who order the universe. An intermediate position was afforded by Charles S. Peirce, the 'father' of semiotics. According to Peirce (1958), the universe consists of things which are real, that is, they exist whether we think about them or not. We experience the real directly. However, our ideas of the real are selective constructions based on our previous experience, history and purpose. It is thus that the mind constructs the order which characterizes our individual perceptions of reality.

The epistemological position in the present work (as in the previous two volumes) is as follows: regardless of whether human beings think about it or not, order *does* exist in the universe; more importantly, the presence of order is a manifestation of a more basic property of the universe, a property which we call ***information***. This position in no way contradicts Peirce's position that, whatever order may exist in nature, our minds may superimpose our own mental order on top of the natural order.

In the discussions which follow, we shall pay only scant attention to these philosophical considerations. The thoughts of great minds, although profound in their time and still of considerable historical interest, were constructed in an age of even greater ignorance than the present. If at present we are still in a woeful state of ignorance, our mindset has, at the least, had the historical experience of uncovering the information code which governs the inheritance of traits in all living organisms: ***the DNA code***. In addition, we have had the shattering experience of working with ***computers*** – artificial devices which mimic aspects of thought processes which go on inside our heads.

The Need for a General Theory of Information

We live in an age of the computer. The technological miracle of electronic intelligence has propelled us into an age which is in the process of becoming as different from the industrial era as that period of our history differed from Europe's Middle Ages (Stonier, 1983). As we move deeper and deeper into this new age, we are creating a global society whose dependence on computers grows exponentially.

We are thus confronted by a dilemma. On the one hand, we will need to rely ever more on computers as we attempt to coordinate the activity of the thousands of millions of people who make up global society. Only with the aid of computers can we create the necessary global infrastructure: air transport, banking, commerce, financial markets, tourism, telecommunications, weather forecasting, etc. On the other hand, this need means that computers will, increasingly and inevitably, take over aspects of judgement and decision making from their human counterparts: from flying an aircraft to sending out severe weather warnings, from forecasting stock prices to diagnosing a patient's illness.

Nowhere is this trend more evident than in the developing military technology. The shift to computer-based command decision systems is necessitated by the increasing speed with which decisions need to be made. The pressure of time can lead to serious human errors. A tragic example is the shooting down of a civilian Iranian airliner by an American warship in the Persian Gulf in July 1988. The nervous commander, responsible for the safety of a ship and its crew, ordered a missile to be launched at an incoming blip on his radar screen before it could reach striking range. If he had had more time, he could have ascertained that it was an airbus – not a fighter – flying on course in a well-defined civilian air lane.

Would an expert system controlling missile launch have behaved differently? As military systems become computerized, and as more and more countries manage to introduce advanced technologies to their armed forces, the matter can no longer be considered as academic. It is true that, usually, elaborate safety backups have been built into such systems – especially those involving strategic nuclear strike forces. However, there are two dilemmas which should give cause for alarm. First, a system to negate launch in response to false alarms needs to be balanced by systems which activate launch before it is too late. Second, systems complexity is subtle: not only do the chances of a malfunction increase exponentially as systems grow more complex, but after a point, complex systems acquire properties which cannot be anticipated.

The mid-1980s saw three high-technology failures – the mid-air explosion of the Space Shuttle *Challenger*, the poisonous fumes of Bhopal descending on a sleeping and unsuspecting city and the Chernobyl meltdown. All three involved major human errors. It is not unnatural that one response to these disasters was to try to reduce the chances of such future mishaps through the use of computer-based expert systems.

Knowledge Engineering

NASA (the National Aeronautics and Space Administration) has always been associated with the most advanced state-of-the-art technology. When it came to the Space Shuttle, its creators considered it to be not only the world's most complex piece of machinery, but also a symbol of American technological ingenuity. This pride was struck a severe blow when the ***Challenger*** blew up shortly after launch. The tragedy brought to mind that other near disaster: the ***Apollo 13*** mission to the Moon. In both cases there were manufacturing design flaws. In the former, a seal failed; in the latter, a thermostat which was supposed to be able to run on both 28 and 65 volts, was only able to handle the lower voltage and created the electrical problems which ultimately blew out the oxygen tank – an almost fatal blow to the ***Odyssey*** command module (Jensen, 1996).

The increasing use of computers for quality control, operations and management means an ever-increasing role for electronic intelligence to augment human judgement. In light of the above, we had better make it work well! A major reason for writing this trilogy, therefore, is to help clarify some of the problems and concepts in a fledgling new discipline – ***knowledge engineering*** – in particular, to provide a theoretical infrastructure.

The professionalization of any engineering discipline tends to exhibit evolutionary processes which share common stages (Gibbs, 1994). Theory usually does not enter a field in any serious way until after the field has had an extensive commercial history and has accumulated a large body of practical experience. Thus, a Galileo appeared only centuries after cannons had been in use; Carnot and thermodynamics developed only after a century of experience with steam engines; and communications theory emerged after the establishment of telegraph and telephone networks. The same may be said of software engineering.

A steam engine, no matter how well designed, is useless if it is not supplied with water for its boilers and fuel to heat the water. Similarly, a computer, no matter how sophisticated, is useless if you have no data to feed into its memory and no rules to drive the inference machinery designed to process such data. Knowledge engineering attempts to codify and distil information so that it can fuel the increasingly powerful logic machines processing the data stored in their memories. Intelligent knowledge-based systems (IKBSs) and expert systems are computer-based systems which take information and process it in an 'intelligent' manner. To do this effectively is the task of the knowledge engineer.

The irony is that whereas today the mechanical engineer working with water and steam understands both the nature of the inputs and the thermodynamic processes going on inside the steam engine, the knowledge engineer has virtually no understanding of the phenomenon that we call 'intelligence', and to confound matters, there does not even exist a genuine theory of information. What generally passes for a theory of information, is, in fact, only a theory of communication. Thus the software engineer struggling to become a knowledge engineer is in the embarrassing situation of understanding neither the true nature of the input (information) nor the true nature of the processes going on inside a computer (intelligent operations).

One problem with creating an ***information science*** is that it requires expertise from a wide variety of traditional disciplines. A number of conferences have attempted to do this. It is worth mentioning a conference held in Madrid in 1994, *Foundations of Information Science*, because its subtitle gives an indication of the scope of the field: *From computers and quantum physics to cells, nervous systems and societies.*

Thus, to achieve a good theoretical grasp of concepts such as information, intelligence and meaning requires an interdisciplinary approach. This leads to a problem – a problem which was described succinctly over forty years ago by Schrödinger (1944, p. vii) in the preface to his classic little book *What is Life?*: 'A scientist is supposed to have a complete and thorough knowledge, at first hand, of *some* subjects, and, therefore, is usually expected not to write on any topic of which he is not a master'. Schrödinger, however, argues that: '...The spread, both in width and depth of the multifarious branches of knowledge... has confronted us with a queer dilemma. We... are only now beginning to acquire reliable material for welding together the sum-total of all that is known into a whole; but, on the other hand, it has become next to impossible for a single mind fully to command more than a small specialized portion of it'. He concludes: 'I can see no escape from this dilemma... than that some of us should venture to embark on a synthesis of facts and theories... at the risk of making fools of ourselves'.

Some Definitions and Problems

Physics could not emerge as a science until it was able to define precisely, and then quantify, phenomena such as *energy*, *work*, *force*, *momentum* and *mass*. The same is true for the construction of a science of information. The problem is that almost all of the terms used to create a general theory of information are common words used in everyday language. These words have not only a ***denotation***, but also a ***connotation***. As such, they may have a different meaning for different people.

Pedro Marijuán, in his opening address to the First Conference on Foundations of Information Science held in Madrid in 1994, pointed out that over the last few decades, the term 'information' has become a 'Jack of all trades'. The term has been applied not only to the theories of the communications engineers (e.g. Shannon), but to the following as well: to all new communications technologies including the mass media; to library science; to hardware and software engineering; to the formulation of the second law of thermodynamics and the concept of entropy; to the contradictions related to the measurement process in quantum theory;

to non-linear dynamics, to the evolution of non-equilibrium systems; to the evolution of living systems; to the concept of ecological diversity; to cellular DNA and to enzymatic processes; to the evolution of the nervous system and to processes going on in the brain; to explaining the nature of intelligence and to the work on artificial intelligence; to the foundations of epistemology and ontology; and to economic, social and political philosophies.

This list is by no means exhaustive. As Marijuán comments: 'The use of the term information in all these provinces is not incorrect, but its overall coherence has become minimal'.

One common perception of the term 'information' is that it is part of a spectrum: 'data', 'information', 'knowledge' and 'wisdom'. ***Data*** is generally considered to be a series of disconnected facts and observations. These may be converted to ***information*** by analysing, cross-referring, selecting, sorting, summarizing, or in some other way organizing the data. Patterns of information, in turn, can be worked up into a coherent body of knowledge. ***Knowledge*** consists of an organized body of information, such information patterns forming the basis of the kinds of insights and judgements which we call ***wisdom***.

The above conceptualization may be made more concrete by a physical analogy (Stonier, 1983): consider spinning fleece into yarn, then weaving the yarn into cloth. The fleece could be considered analogous to data, the yarn to information and the cloth to knowledge. Cutting and sewing the cloth into a useful garment is analogous to creating insight and judgement. This analogy emphasizes two important points: (1) going from fleece to garment involves, at each step, an input of work; and (2) at each step, this input of work leads to an increase in organization, thereby producing a hierarchy of organization.

Although, in the present work, the term ***information*** has a very different conceptualization from the more common one presented above, the common conceptualization does intuitively imply that there are *hierarchies*, and that to get from one level to the next requires an input of work. A general theory of information must address this question of the origin and nature of information hierarchies.

Furthermore, a general theory of information should be able to define precisely terms and concepts such as *message*, *misinformation*, *negative information* and '*structural*' and '*kinetic*' *information*. Some terms, such as *intelligence*, *consciousness* and *self-consciousness* have already been explored extensively in the second book of this series: *Beyond Information: The Natural History of Intelligence*. Others, such as ***meaning*** will be considered later in the present volume, as will the phenomenon of ***consciousness***.

Likewise, a general theory of information should address itself to the general ***properties*** of information and be able to characterize basic ***processes***, such as how information is created, how it is processed, how it is transmitted, how it is detected, how it is interpreted and how it is stored and retrieved.

Until Recently, Why Has Information Been Ignored?

If information has a physical reality, as does energy, why has it been ignored for so long? There are two interrelated reasons why this has happened. The first is that until the second quarter of this century there was no ***need*** for a concept such as 'information'. It was only after the advent of complex telephone networks, and, later, computers, that the necessity for considering information as an abstract entity in its own right began to be appreciated by the engineering professions.

The matter parallels the history of the concept of 'energy'. Prior to the 16th century there was no great need to develop such a concept. A body was hot or cold in the same way that a body might be hard or soft. 'Heat' was an inherent property of a body which might change under various conditions, the way a stick of butter might be hard or soft. Of course, there was such a thing as ***force***: the strength of a person pushing, or an ox pulling, or, of increasing importance, the strength of the wind pushing the arms of a windmill, or the force of rushing water propelling a boat or an underslung waterwheel. Most dramatic was the force imparted to a cannonball by exploding gunpowder. It was this phenomenon which preoccupied Galileo during his career as a military engineer. Galileo, as we all know, rolled various sized cannon balls down an incline to discover the laws of trajectories. Between conducting experiments with cannons and observing the motion of planets – coupled to the ferment of the time, it was not unexpected that the 'Father of Physics' laid the groundwork for the concept of ***energy*** – a concept which, however, did not reach its full maturity until the 19th century, when Joule was able to quantify the interconversion of one form of energy into another.

In like manner, it was the telephone engineers, such as Hartley and Shannon, who started to look at information as an independent, if abstract, quantity. Their concern was with the ***transport*** of information – specifically, how much information could be moved from sender to receiver via a noisy channel. However, it was the advent of information-processing machines – computers – which really provided the impetus for considering information as something both abstract and real – a somewhat intangible entity which somehow existed in its own right.

In the meantime, new discoveries and the further intellectual evolution in both the biological and physical sciences, set the stage for broadening our ideas about information as a phenomenon. Initially, biologists were confronted with the problem of genetic information: how is the information which results in the colour of the eyes, or the shape of a limb, stored and transmitted from one generation to the next? The deciphering of the DNA code – among the most profound of human discoveries – made it clear that there existed something distinct from 'human' information, something much more ancient. The concept of bio-information became enlarged over the next few decades as it became clear that the cell is a highly sophisticated information processing device and that even individual molecules such as enzymes,[1] by virtue of their structural organization, must possess information in order to operate selectively.

Likewise, some physicists have started to look at information differently. In 1929, Leo Szilard published a paper on Maxwell's Demon, in which he explored the idea that entropy and information are interrelated (Szilard, 1929). In 1944, Erwin Schrödinger, whose work we will discuss in the next chapter, presented the idea that negative entropy and order might be associated with living systems. Cyberneticists, such as Norbert Wiener, pointed out (Wiener, 1948, p. 132): 'Information is information, not matter or energy' and elaborated further (Wiener, 1961, p. 11): 'The notion of the amount of information attaches itself very naturally to a classical notion in statistical mechanics: that of ***entropy***. Just as the amount of information in a system is a measure of its degree of organization, so the entropy of a system is a measure of its degree of disorganization; and the one is simply the negative of the other' [*italics in original text*]. The study of open, dissipative systems, chaos theory and, more recently, complexity theory is beginning to lay the groundwork for a new conceptualization of the role of information in physical systems.

Recently, astrophysicist David Layzer (1990, p. 265) has argued that: '...order, although seemingly less concrete than matter and energy, is just as fundamental a constituent of the external world'. He defines order as the opposite of randomness. Unfortunately, Layzer does not take the next step, i.e. that order is a reflection of the information content of the system. This unwillingness to invoke information as a basic principle reflects the present-day culture of physics. The main impediment to a full

---

1 Enzymes are organic catalysts which, like all catalysts, speed up chemical reactions without being used up themselves.

acceptance by the physical scientists of the hypothesis that information has as much reality as matter and energy is the perception that the physical sciences do not need such a concept. As long as accounting devices such as ***potential energy***, ***latent heat*** and ***entropy changes*** are considered to be a satisfactory explanation of phenomena which involve changes in the information status of a system, physical scientists and engineers will ignore the reality of information.

·2·

The Nature of Information

What is Information?

Information is one of those terms which everybody thinks they understand, but the more closely we scrutinize the idea, the more elusive it becomes. A dictionary may define 'information' in terms of 'facts', 'knowledge', or 'experience', but then defines these terms by referring them to each other, or back to information. It is clear that, although we all have an intuitive understanding of the term 'information', our understanding is not sufficient to allow us create, for example, a theory of information which would allow us to explain manifestations such as meaning, knowledge, insight, or wisdom. Nor, at the other end, when we are dealing with the smallest units of information – bits of data – do we have a satisfactory theory which can explain how these bits interact to create comprehensible language.

What is information? One of the best answers to that question comes from Lila Gatlin's insightful book Information Theory and the Living System: 'To be honest, information is an ultimately indefinable or intuitive first principle, like energy, whose precise definition always somehow seems to slip through our fingers like a shadow'. (Gatlin, 1972, p. 25).

Gatlin's comparison to the concept of energy is fortunate. As we shall see, there are many parallels between information and energy. For example, as we already discussed, historically energy did not become a scientific preoccupation until it began to significantly exceed humanity's experience with nature – as it did with the firing of cannons. Similarly with information: it was only after telephone networks had become commonplace that the communications engineers began wrestling with the question of how much information one could transport through a telephone system. More recently, of course, the advent of the computer has created a whole host of new questions about the nature of information.

Information and the Laws of Physics

One of the main themes of the present series on information revolves around the idea that the description of all physical systems entails not only the parameters which define the amounts of matter and energy, but also the quantity of information. Furthermore, any changes in the systems must take into account not only changes in matter and energy, but also changes in the information content of the system. That is, just as we ascribe to matter the mass encountered in our universe, and to energy the heat and other forms of energy, so must we ascribe to ***information*** the organization (or lack of it) which we encounter in all systems.

The idea that information is an intrinsic component of all physical systems requires a reevaluation of the laws of physics, including their fundamental equations (all of which contain, intrinsically, information statements). Among these laws and equations are those which define thermodynamics. Tampering with the laws of thermodynamics is, of course, a dangerous business. As Sir Arthur Eddington admonished: '...if your theory is found to be against the second law of thermodynamics I can give you no hope; there is nothing for it but to collapse in deepest humiliation' (Leff and Rex, 1990, p. viii).

Biologists are not so awed by this second law, obviously sacrosanct to physicists. Biologists, in their realm, see it constantly reversed. True, this would not happen without an input of energy from the Sun – but there is more to the phenomenon of ***life*** than that its entropy is tending to decrease at the expense of the Solar System. The question posed by the existence of biosystems is this: why and how do there exist systems whose entropy tends to decrease with time? A number of physicists have also been puzzled by this question, most notably Erwin Schrödinger. It was Schrödinger who pointed out that: 'living matter, while not eluding the "laws of physics" as established to date, is likely to involve "other laws of physics" hitherto unknown' (Schrödinger, 1944, p. 69). It is hoped that the theory presented here and in earlier works by the author, will help to clarify what some of these new laws might be, and how the existence of information as a physical parameter leads to a reinterpretation of established theory.

Shannon's Theory of Communication

The traditional view, shared by most physical scientists and engineers, in which ***information*** is considered to be an abstract entity, is usually credited to Claude E. Shannon (1948). This work was restated and popularized by Shannon and Weaver (1949), although, in actuality, Shannon never claimed to have developed a theory of information. Instead, Shannon considered his contribution to have been a theory of communication – i.e. a theory of information ***transport***. Earlier, in 1928, R.V.L. Hartley (as reviewed by Cherry (1978)) had suggested a way of quantifying the 'information' contained in a message. The suggestion represents one of those events in history where a whole field is switched onto the wrong track. As Colin Cherry has commented: 'it is a pity that the mathematical concepts stemming from Hartley have been called "*information*" at all' (Cherry, 1978, p. 51). Cherry goes on to point out that the formula derived by Shannon for the average information contained in a long series of symbols is really a measure of the statistical rarity or 'surprise value' of a course of message signs. This is hardly a true measure of the information content of a message.

To confuse matters further, Shannon introduced the concept of entropy into his calculations. The idea that ***entropy*** and information are somehow interrelated had been around for two decades. It had been postulated by Szilard (1929). However, Shannon's basis for doing so stemmed from the fact that his equations ***looked like*** the equations that had been developed in statistical mechanics by the Austrian physicist Ludwig Boltzmann.

It is said that he was also encouraged to do so by Von Neumann, who told Shannon that since '...no one knows what entropy is, so in a debate you will always have the advantage' (Campbell, 1982, p.32). One could argue that, as a result of Von Neumann's advice, the communications engineers and information theorists all became the victims of a bad joke: that the potential indeterminacy of a message is the same thing as entropy. The confusion still reigns today.

Shannon's sleight of hand has been attacked by a number of authorities: Yockey (1992), after examining the matter carefully, concludes that: '...there is, therefore, no relation between Maxwell–Boltzmann–Gibbs entropy of statistical mechanics and the Shannon entropy of communication systems'. Likewise, Wicken (1987. p. 179) has shown that: 'While the Shannon equation is symbolically isomorphic with the Boltzmann equation, the meanings of the respective equations bear little in common'. Citing the work of Brush (1983), Wicken points out that both Boltzmann and Shannon were independently using equations describing games of chance – equations which had already been developed two centuries earlier by French mathematicians. A comprehensive critique of Shannon's conceptualization (not his mathematics) has been presented earlier by the author (Stonier, 1990b).

Information and Entropy

The concept of ***entropy*** emerged following a century of experience with the steam engine. The concept was implicit in the work of Rudolph Clausius in 1850; however, he did not suggest the term entropy (derived from the Greek expression for ***transformation***) until 1865 (Tisza, 1991). Initially, the problem was to account for the unidirectional dissipation of work into heat, and how to quantify this apparent loss of energy. Subsequently, the problem was clarified by the work of Lord Kelvin and extended by Gibbs, Boltzmann and others.

It was the explorations in statistical mechanics by Boltzmann (1896, 1898) which helped to connect changes in entropy to changes in the order/disorder of the system under study. It was Boltzmann's equation which also led Schrödinger (1944) to consider the implications of entropy and order for living organisms.

Schrödinger's speculations made it possible to view ***information*** from an entirely different theoretical perspective. Although Schrödinger did not write it as an equation, it is clear from the text (Schrödinger, 1944, p. 73) that he considered 'order' to be the inverse of 'disorder'.

The information theory proposed earlier by the present author (Stonier, 1990b), assumes information to be a basic property of the universe – as fundamental as matter and energy. Specifically, a system may be said to contain information if such a system exhibits organization. That is, just as mass is a reflection of a system containing matter, and heat is a reflection of a system containing energy, so is organization the physical expression of a system containing information. By 'organization' is meant the existence of a non-random pattern of particles and energy fields, or more generally, the sub-units comprising any system.

It is important to differentiate between the concept 'information' and the concept 'message'. Most lay people and many, perhaps most, experts do not consider an organized pattern to contain information until it contains a message. This is in contrast to how 'information' is defined here.

Information is the raw material which, when ***information-processed***, may yield a message. It does not matter whether this raw material consists of the letters of a written alphabet, the nucleotides of a strand of DNA, the atoms making up a crystal, or patterns of electromagnetic radiation. They become a message only after the raw material (information) has been information-processed. Upon receipt of a message, the message must once more be information-processed by the recipient in order for the message to acquire ***meaning***.

Without this precise differentiation between *information, message* and *meaning* there can be no theory of information. Without this differentiation it becomes impossible to quantify the information content of a system.

Each individual letter, nucleotide, atom, or electromagnetic pulse itself contains information. All these entities exhibit some form of organization (the last in terms of wavelength and amplitude). This means that two moles of sodium chloride will contain twice as much information as one mole, since two moles will contain twice as many molecules. Similarly, two moles of DNA will contain twice as much information as one mole. However, if the nucleotide sequence in the second mole is identical with the first, then the second mole would convey a message which is merely redundant. The same may be said of two identical sentences, one following the other: the two sentences would contain twice as much information as the single sentence. However, sent as a message, the meaning of two such redundant sentences would not double, although it might be increased somewhat as a result of being repeated.

In order to spare the reader who is not inclined to follow mathematical arguments, all calculations have been banished to Appendix A at the rear of this book. What the proposed theory concluded was that the relationship between information and entropy was quite different from that understood by the communications engineers.

Furthermore, **entropy**, this strange 'state of a system', must not be viewed as a simple *metaphor* for describing information. Changes in entropy actually do measure the changes in the information content of physical systems: an increase in entropy measures an increase in the ***absence*** of information. This is demonstrated most clearly by observing what happens when we heat ice: when the ice melts and loses its structure, there occurs a great increase in the entropy of the system. Thus, such a great increase in entropy occurs when, before our very eyes, we observe a change in the organization of matter. The same occurs when we continue to heat the water to boiling. Once again, as the boiling point is reached, there is a massive rise in entropy (Figure 2.1, *overleaf*).

This phenomenon occurs repeatedly at every phase transition (Greiner and Stocker, 1985). As we continue to heat the steam there comes a point where the molecules ionize and then break up into atoms, which after becoming stripped of their electrons may be broken up into nucleons, which, given enough of an energy input, can be further fragmented into quarks and gluons. At each such phase transition we observe great increases in entropy. But it is also at these phase transitions that we encounter a dramatic destruction of bonds and the disorganization of matter. By definition, a destruction of organization reflects a reduction in the

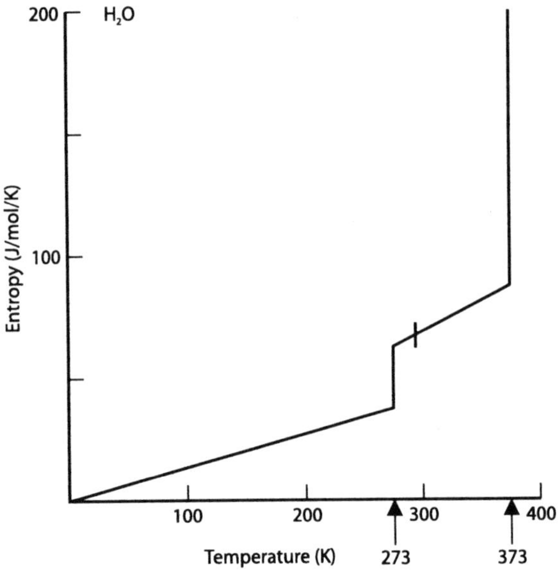

Figure 2.1 Plotting entropy against temperature

information content of the system. Hence, an increase in entropy and a corresponding decrease in the information content of a system go hand in hand.

If these postulates are correct, one should be able to calculate a conversion factor. By examining the entropy changes which occur in two different systems it was possible to ascertain the changes in information states and derive a general equation relating entropy units to information (Stonier, 1990b). Specifically, the calculations indicated that one entropy unit equals approximately 10^{23} bits/mole.

What this means is that, depending on the temperature involved, small amounts of energy can, in theory, be converted into very large amounts of information. This parallels the well known equation relating ***mass*** to energy, which shows that very small amounts of matter can create very large amounts of energy.

The Interconversion of Information and Energy

Every time I pick up a pencil from the floor and place it on the desk, I expend energy to alter the organization of the universe. In placing the pencil on the desk, I have engaged in work; I have also created a thermodynamically less probable situation – hence I have ***increased*** the information content of the universe. Traditionally, it always has been a mystery as to what happened to the energy expended when one picks up a pencil. To explain it, physicists were forced to invent an accounting device: ***potential*** energy.

Potential energy has always been an anomaly. The pencil on the desk is devoid of energy. It will not move until I apply another force. The present theory proposes a much simpler explanation: the energy expended became converted to information. The loss of energy equalled the increase in information of the system.

To emphasize this point: ***work*** is a transient phenomenon; on the other hand, the ***product of work*** involves a change in the information status of a system, a change which remains fixed until such time as it is subjected to another force or to further work. In line with this conceptualization one may define ***potential energy*** as: **a term which describes a *state* in which the *expenditure of energy* has resulted in an *increase in the information content* of the system.**

Information as a Fundamental Property of the Universe

The implications of some of the postulates for thermodynamics and for physics, in general, have been dealt with in the earlier work (Stonier, 1990b). However, it is appropriate to review certain aspects at this point, since ***information***, as a fundamental property of the universe, has many interesting properties of its own. Many of these are analogous to the properties exhibited by energy.

1. Energy has, as one of its fundamental attributes, the capacity to perform work. Similarly, *information has, as one of its fundamental attributes, the capacity to organize things.*
2. Like energy, *information, in some form, is embedded in every equation of the physical sciences.* Information is present in the form of time, distance, or direction (velocity, acceleration), or in terms of some constant. Such constants could not exist if the system were not organized.
3. *Like energy, information may be added to matter. The former causes matter to heat up; the latter causes matter to become organized.*
4. *Information may also be added to energy, causing such energy to become of a higher grade.* All forms of energy except heat exhibit distinct organization and therefore contain some information as

well. Only heat, in the sense of the uncorrelated motion of particles or quasi-particles (e.g. phonons), contains no information.

5. *Information may also be added to itself, resulting in higher levels of self-organization.* The phenomenon of information hierarchies has already been alluded to above and will be discussed in greater depth in a later section.

6. *Energy and matter are* ***interconvertible****. So are information and energy.* The law of the conservation of energy is basic to contemporary thermodynamics. However, the *law of the conservation of energy* needs to be expanded to read: in a closed system, what is conserved is the sum total of energy plus matter plus information. Energy may be converted into either matter or information. The former is defined by the equation $E=mc^2$; the latter is defined by the equation $E=IT$, where E is measured in joules, I is measured in information units (approx. 10^{23} bits) and T in kelvins.

7. The continuous addition of heat energy to matter results in reaching certain critical values, at which points there occur ***phase transitions***. Likewise, *the continuous addition of information to matter results in reaching the equivalent of a phase transition in which systems become reorganized into new kinds of patterns.* The phase shifts associated with the addition of information result in the creation of hierarchies of information. Such hierarchies span a spectrum of phenomena ranging from the sub-atomic to human societies and their abstract creations.

8. *Like energy, information may exist in many forms.* Energy may manifest itself as heat, light, sound, electrical, chemical, osmotic, atomic, etc. forms of energy. Information may manifest itself as structural, kinetic, temporal, spatial, biological, human linguistic, machine coded, etc. forms of information. Two major classes of information are ***structural*** and ***kinetic***. A steam engine in action exhibits both. First, the engine itself has embedded in it a history of information inputs. This exemplifies structural information: information which will not change (except for wear and tear) during the operation of the engine. Second, a steam engine acts like a Maxwell's demon: it arranges water molecules in such a fashion that the high-energy molecules end up on one side of the piston, while the low-energy ones end up on the other. This non-random distribution represents information – however, it is transient: it is information about to be put to work. As the potential energy is converted to kinetic energy, the ***kinetic information*** declines in parallel with the potential energy.

9. Like energy, information may manifest itself in grades. In more traditional terms, mechanical energy, chemical energy and heat represent high-, intermediate- and low-grade energy. During work, the degradation of higher grades of energy into (low-grade) heat is measured by an increase in entropy. Entropy measures the loss of information associated with the energy which is performing the work. Sub-nuclear atomic organization may involve the most simple sort of structural information. Adding information to information, as with adding information to energy, results in higher grades of each. Information hierarchies reflect grades of such information.

10. *The information contained by a system is a function of the linkages binding simpler into more complex units.* There exist many different kinds of linkage.

11. *The term **information** is conceptually different from the term **pattern**.* Information is the abstract quantity which represents the common coin. This may be best illustrated by looking at the parallels between information and energy. For example, heat, light and electricity are different from each other. Yet we may obtain light in a light bulb from electricity generated by a power plant burning fuel to produce the heat to drive the generators. Light, electricity, heat and mechanical energy are all forms of energy: ***energy*** is the abstract quantity which represents the common coin. In like manner, a ***pattern*** of dye molecules on the printed page of a book is very different from a ***pattern*** of compressed air molecules carrying the voice of a person reading out loud. Yet the information transmitted by both may convey the same meaning. Likewise, the pattern of neural connections which sends out an alarm signal to the body is quite different from the patterns of atoms on the molecule of adrenaline which activates the various cells, tissues and organs into a response. However, the message being transmitted. ‘Danger get ready!’ is the same in both cases. ***Information*** is the common, convertible coin.

12. *Just as energy may exist in particulate form (photons), so may information (infons).* Photons are devoid of mass but exhibit relativistic momentum when travelling at *c* (the speed of light). Infons would travel at velocities other than *c* and lack both mass and momentum: they would contain only information. The migrating ‘holes’ left behind by migrating electrons in a semiconductor would exemplify such particles.

Infons

The idea that physical information may exist in particulate form is conjectural. Detecting infons, if they exist, would be even more problematical than detecting neutrinos. Nevertheless, it is worth spending a little time on this concept because the idea that information is a basic property of the universe becomes more real if one recognizes the possibility that information actually may exist in the form of physical particles. Once the idea of this possibility is acknowledged, the paradigm shift required of information theorists in order to develop a general theory of information becomes complete. For if it *is* possible for information to exist as physical particles – pure raw information, as it were – then the conceptual linkage which insists that ***information*** must be tied to a ***message*** (which must have ***meaning***), can be broken.

One of the best examples of information behaving as if it consists of particles is found in biological information systems. In the middle of the 19th century, the Augustinian monk Gregor Mendel discovered that certain inherited traits in the common garden pea segregated out as distinct units. These units of information, which determined whether the next generation grew tall or dwarf, or whether the pea was wrinkled or smooth, behaved as if they had an integrity of their own – in other words, they behaved in particulate fashion.

Subsequently, these 'units of inheritance' were given the name 'genes', that is, they were considered as particles of genetic information. As we now know, genes are made up of nucleic acids, which, from a molecular point of view, means that genes are large structures. From a human perspective, however, genes are particles that are so small that not only do we have no hope of seeing them with the naked eye, we cannot even detect them with the most powerful of optical microscopes. Biological information, therefore, may exist in the form of invisible particles. Genes, therefore, may be considered to be the biological equivalent of infons, keeping in mind that biosystems are much further up the hierarchy of information-processing systems than are physical systems.

Let us now turn to such ***physical*** systems and explore the possibility that information may exist within them as particles. One of the features which made Times Square in New York a centre of attraction was the electric sign alongside the Times Building. The latest news bulletins flashed along the sign, moving like a gigantic tickertape – bringing, over the years, entertainment and enlightenment to millions of passers-by.

As the letters and words migrated along the sign, the illusion was created that the light bulbs were moving across the screen as discrete entities. The light bulbs, of course, didn't move at all. They were completely stationary. However, because they represented binary information devices, their on/off states allowed patterns of light to traverse the length of the sign.

Such patterns of light, moving across a fixed distance at a fixed rate, exhibit both frequency and wavelength – easily discernible if, instead of words, just a series of bars or upper-case letters 'I', were moving across the sign. Such waves do not represent waves of energy: the energy is confined to the flow of electricity within the circuit of the light bulb. The moving patterns do, however, represent waves of information.

Our universe is filled with examples in which changes in patterns of organization – by definition, changes in information states – result in the appearance of waves or particles. The emission of a photon by an atom as an electron drops from an outer shell to an inner one is one such example. Another example results from the complex interaction of waves resulting in a phase wave. This example is interesting because it represents information superimposed on a carrier wave. Such a wave, superimposed on a light wave, could, in theory, move faster (or slower) than the electromagnetic wave which carries it. That is, in theory, such an information wave could move faster than the speed of light.

Let us return to Times Square. If not the New York Times sign, there exist signs in which the letters and symbols moving across the field of lighted bulbs are dark. That is, each letter is created by a pattern of electric bulbs which are 'off'. These 'off' states appear as dark 'holes' moving across the brightly lit field. Each binary light bulb, therefore, represents not only a discrete particle of information, but the information is conveyed to the passer-by by means of patterns of moving 'holes'.

Inside a crystal, 'holes' may appear and move as if they were particles. As with the Times Square sign, the holes don't actually move. It is the sequential opening up and filling of holes which produces the *illusion* that the holes are moving. Although it requires energy in the form of applied forces for such a phenomenon to occur, the apparent movement of the holes represents a movement in the *organization* of the crystal. It represents information on the move. Thus, massless particles like phonons and excitons, thought to be important in explaining the behaviour of electrons in superconductivity states, could be viewed as representing particles of information. Like the light bulbs on the Times Square sign, the behaviour of phonons and excitons represents a particle-like manifestation of gross changes in organization, changes which appear as patterns of information flows across the system.

In physical systems, unlike the biological one described above, infons would be noted for their ***absence*** of structure, rather than for their positive nature (as, for example, a gene). Holes, or empty spaces, when enclosed in an organized structure, may be highly significant pieces of information. To understand the importance of the absence of something in the context of some structure or some information system we might consider the following three examples: the rooms in a house, the number zero and the space between letters.

1. When we build a house, we build it to enclose empty spaces which we call rooms. It is true that we may wish to fill up these spaces with furniture and appliances, but the importance of these spaces as spaces is demonstrated by the fact that when we rent or buy a house we ask how many bedrooms, how many bathrooms, is there space for a den, and only secondarily (if at all) how many feet long and wide is the house? The rooms, which are our major preoccupation, are empty spaces within the confines of an overall structure: the house.
2. The history of mathematics is a history of the capacity of our human species to engage in abstract reasoning. For numbers to be conceived of as *numbers* – that is, as entities in their own right – involved climbing up a ladder of abstraction. The experiences of sectioning off fields, of building temples and fortifications, of engineering aqueducts, of studying the motion of planets – all these led to other lines of abstraction which became embodied in geometry and trigonometry. However, for the ancient Greeks, the Romans and European scholars right up to the Middle Ages, numbers were an unsolved problem. Only when the number zero, initially invented in India, was introduced by the Arabs into Europe as part of what became known as Arabic numerals, did European mathematicians such as Fibonacci begin to work with numbers effectively. The number zero, of course, represents the ***absence*** of a number.
3. The 'Latin' alphabet in use today was derived from the Romans, who borrowed it from the Etruscans, who, in turn, had taken it from the Greeks around 700 BC (Starr, 1991. p. 449). However, the use of the alphabet during most of the first millennium was very different from how it came to be used in post-medieval times (Illich and Sanders, 1989). For example, Latin texts prior to the 7th century lacked spaces between words – the letters simply ran on. Dividing sentences up into words was a pedagogic device introduced in the 7th century for the benefit of 'ignorant' Celts preparing for the priesthood. Breaking up sentences into words was a means of teaching Latin as a second language. However, it soon became apparent that this was a most useful device – by the 9th century, schools

> around Europe had begun to accept separation into words as a matter of course. Thus there appeared the space, a symbol signifying '***no*** letter'.
>
> In fact, the most commonly used symbol in the written English language is the space between words. On a typewriter, the space bar is used more frequently than any other single key. Paradoxically, the information content of the space (the hole) between two words disappears the moment the words on either side disappear and one is left with just a blank page.

Thus we must recognize as axiomatic two statements: (1) the ***absence*** of structure within a structure may carry information as real as the structure itself. Holes and spaces within an organized structure may comprise a significant part of the organization of that structure, and hence contain information. (2) The information content of such holes or spaces is entirely dependent upon the organization and behaviour of the structures or systems which surround them. This means that there exists information which can exist only as long as there exists a context or structure – a form of information which appears to disappear the moment the structure disappears.

Atomic shells are highly structured entities. If they weren't, neither Balmer's ladder, nor Pauli's exclusion principle, nor any number of other atomic phenomena would manifest themselves. When a photon is absorbed by an atom, its energy manifests itself by ejecting an electron from an inner shell. At least part (and perhaps all) of the photon's information may manifest itself as the hole left in the inner shell which lost the electron. That is, it would not be unreasonable to postulate that the hole left in an atomic shell by a lost electron, constitutes a particulate form of information, i.e. an infon.

Such holes are known to move across crystalline structures in collective excitations with a velocity, frequency and wavelength. In themselves they possess neither mass nor momentum, and their velocity is not confined to *c*. As such, they would fulfil all the requirements of 'infons' postulated above.

Holes comprising particulate information (infons) inside organized structures need not be confined only to structures of matter, such as a crystal. Such holes might well exist inside fields of organized energy, such as a gravitational or magnetic field. Furthermore, there is no reason to believe that such holes could not migrate across such fields as they do across material structures. In other words, infons could move not only inside matter, but across force fields as well: if the context provided by the organization of the enveloping atoms is removed, there remains the possibility that fields of gravity or magnetism, or other force fields, could also provide a matrix for such a moving particle of information.

It is interesting to note that, if the relativity equations are interpreted in a certain way, they contain within them the possibility of the existence of such particles (see Appendix B). Such 'infons' might not show up in traditional physical experiments, since such particles would possess neither mass nor energy. As such, they would not interact with any detectors which are based on the interaction of an energetic particle with atoms of matter. Instead, they might manifest their effect by bringing about changes in the organization of the matrix within which they are operating.

It may well turn out that there exist ***no*** such entities as infons. On the other hand, should the above conjectures prove to be correct, then to ***fermions*** and ***bosons*** we would need to add a third class of particles: ***infons***. The first two represent the particulate manifestations of matter and energy. Infons would represent the particulate manifestation of information, and would include virtual particles such as phonons, excitons and the holes left in atomic shells by ejected electrons. At this point in the historical development of a science of information the possibility of the existence of infons must be left open.

Devlin's 'Infons'

Originally, the term 'infon' was coined by the present author to describe hypothetical particles which, although they lack both mass and energy, possess a ***physical*** existence (Stonier, 1990b). That is, they were considered to be ***real*** particles – 'real' in the sense in which Peirce would have used the term, i.e. they exist whether we think about them or not. Since that time, the mathematician Keith Devlin (1991) has also used the term, but unfortunately, to mean something quite different. Devlin's interesting book *Logic and Information* describes infons on p. 48 as : 'semantic objects within the framework of our theory... their status as "abstract objects" does not mean that infons have any kind of physical existence. An intentional, abstract object is not the same as being a part of the physical world'. In trying to develop a mathematics of information, Devlin invented a category of abstract entities which he has called 'infons', entities which would have the same status as, for example, the 'real numbers' do in mathematics.

In so far as Devlin's notation and name 'are intended to emphasize the fact that infons are semantic objects, not syntactic representations'

(Devlin, 1991, p. 23), one wonders whether the term 'semions' would not be more appropriate, since as Devlin's 'infons' represent the fundamental units of a message (and themselves constitute a message). In the absence of such a change in terminology one can only hope that this second use of the term 'infon' will not lead to the kind of confusion which we have seen over the last half century brought about by the use of the term 'entropy' by earlier information theorists.

The Evolution of the Universe

It may be coincidental, but a graph of Equation A.7 (Appendix A), plotting the increase in information against the decrease in entropy (Figure 2.2), can also be interpreted to represent the increase in information that must have occurred as our universe evolved.

At the far right – where entropy approaches the infinite, and information the zero state – we approach the condition characterized by the Big Bang. As we move to the left, the information content of the universe begins to increase. First the *forces* of nature differentiate out to begin the process of organizing the universe – gravity, weak and strong nuclear forces, electromagnetism – then matter appears. As we move further to the left, we see the evolution of matter into increasingly complex forms. By the time we approach the ordinate – the zero entropy state – self-organizing systems begin to appear.

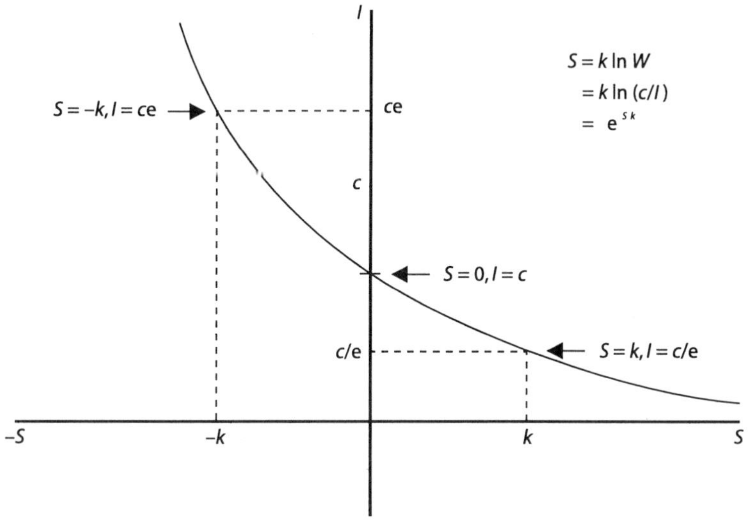

Figure 2.2 Plotting information (I) against entropy (S).

The zero entropy state traditionally has been defined as the lower limit of entropy. Nernst's 'Third Law', generalized by Max Planck (Cohen, 1996) – 'The entropy of a pure substance in a perfect crystalline form is zero at absolute zero' – implies that entropy cannot become negative.

Entropy may be lowered by one of two means: either heat is withdrawn from the system so as to lower the temperature, or information is added to increase its organization. Achieving zero entropy at absolute zero means that virtually no further reduction of entropy can be achieved by lowering the temperature. However, it does not preclude the possibility of lowering the entropy further by adding more information. Returning to Planck's perfect crystal at zero degrees: what might contain less entropy than a perfect crystal of DNA? *Answer*: the naked DNA enfolded in a protein coat to create a functional virus.

Thus, absolute entropy may become negative. As we move into the left quadrant we see not only the further development of more advanced self-organizing systems, we have now reached the realm of self-reproducing systems: i.e. biological systems. We also see the emergence of an entirely new phenomenon – intelligence. From here on in, the curve depicting the growth of information becomes increasingly steeper, reflecting the autocatalytic processes which characterize advanced systems capable not only of organizing themselves, but, with increasing efficiency, also of ordering their environment.

In our part of the universe this process has culminated in human global society. From the perspective of the laws of physics (as they are known at present), the existence of human society is a phenomenon thermodynamically so improbable as to defy imagination. Such improbability of advanced information systems causes one to wonder how it is possible that such systems arose in the first place. The answer lies in the recursive properties of information systems. Organized systems exhibit resonances. Resonances lead to oscillations. Oscillations represent timed cycles during which changes may be introduced. Such changes may dampen or amplify the existing oscillations. Alternatively, they may create new resonances and excite new sets of oscillations. The more complex the system, the greater the likelihood of introducing changes into the system during any given cycle and making it more complex still.

The Evolution of Biosystems

The evolution of biosystems illustrates these principles. If the origin of life still remains a matter of conjecture, it is clear that once cellular life had emerged, biosystems depended on numerous chemical feedback loops for their normal functioning and survival. In due course such feedback loops not only involved the chemical pathways within an individual organism, but involved different kinds of feedback loop (no longer based merely on chemical interactions) between organisms, thereby forming the complex feedback loops which characterize ecosystems.

It took well over half a billion years to evolve from single-celled animals to human beings. It may have taken the previous billion years to evolve the 'eukaryotic' cells of modern single-celled animals. Such cells contain a nucleus, mitochondria and a variety of other cell organelles – all linked into a synchronized whole. At least some, and possibly all, of these organelles were derived from previous life forms at some previous age. These would have included 'prokaryotic' cells lacking nuclei, cells which still cover the globe today in the form of bacteria.

Multicellular organisms arose by linking cells into coherent functioning units. This could be achieved only after the emergence of eukaryotic cells. Prokaryotic cells link only weakly and are able to create only the most simple of structures, such as chains, filaments and clumps (as today do bacteria, particularly the cyanobacteria). For advanced, complex multicellular organisms the eukaryotic cell was a prerequisite. The structure of the cell wall and other membranes allowed for cell attachment and communication between cells in a manner not possible for the more primitive prokaryotic cell. In addition, the complexity of eukaryotic cells allowed for the diversity needed to allow differentiation into a wide range of cell types, and to permit the development of mechanisms for cell migration and attachment.

It would appear to be axiomatic that for eukaryotic cells to have evolved in the first place required the previous existence of prokaryotic cells and other primitive life forms. Early, archaic life forms depended on the prior existence of complex molecules, derived from combinations of simpler molecules, which arose from the forces linking up atoms, which in turn were formed by the intra-atomic forces linking fundamental particles into nucleons and atoms. Complexity utilizes pre-existing complexity to achieve higher degrees of complexity, building up the information content of evolving systems *ad infinitum*. Once this principle is understood, it becomes apparent not only that complex systems evolve, but also why complexity tends to increase at an exponential rate.

Once the eukaryotic cell had evolved to the point of favouring the development of multicellular organisms, a profusion of new life forms evolved. Initially this led to organisms still seen today as the larvae of certain marine invertebrates – microscopic forms consisting of only a few thousand cells (Davidson *et al.*, 1995). This occurred long before the fossil record began – a time during which the increasing complexity of life forms must have favoured the introduction of novel variations in metabolism and in body plan. In due course, entirely new batteries of genes (consisting of mutations, of novel combinations of genes, or of genes acquired from other organisms) evolved to create new patterns of development – patterns which became superimposed, which led to the emergence of much larger and more complex body plans – forms which we now see abounding in the oceans as adult forms. At some point, a phase transition in the evolution of biosystems occurred: the emergence of increasingly larger and neurologically more sophisticated (i.e. smarter) predators. Now the selective pressures on the existing life forms developed a new emphasis: for predators, the increasing sophistication required to find, capture and subdue prey; for prey, the capacity for eluding predators or defending themselves against attack. With the exception of the species at the very top of the food chain, all others now had the problem of not only getting enough to eat, but also of not being eaten. This profound change in the nature of the marine ecosystem vastly accelerated the evolutionary process. The profusion of life forms which appeared at this time (530 million years ago) is referred to as the ***Cambrian*** explosion.

Animals became larger and more sophisticated, and exhibited entirely new features. For example, to defend themselves some animals clad themselves in mineral shells. Others developed hardened spines. Still others just became bigger, and therefore harder to swallow. These larger organisms now might leave tracks in the mud – all of these innovations helped to create a fossil record. As Eric Davidson and his colleagues point out in their closely reasoned paper: '... the Cambrian explosion would have to be regarded at least partly as an explosion not of novel metazoan phyla but of fossils' (Davidson *et al.*, 1995). The progenitors of the various phyla must have existed earlier, and probably for a long time. However, the Cambrian must be considered as one of those periods in the history of ***life*** when there occurred a phase transition, a transition brought about by the advances in the nature of the biosystems such that, once again, the life forms profoundly altered their ecosystem.

Later, the invasion of the land habitat, first by microorganisms, then by plants, and finally by insects and other animals, favoured the further divergence of life forms. Given enough time and a steady input of energy, that totally improbable event, the emergence of higher intelligence and consciousness, was inevitable. As to time – there was plenty of it: compare a century of human affairs with a billion years of geologic time. As to energy – the warming rays of the Sun (neither too close nor too far away) provided the steady input needed to drive the whole process, allowing improbability to feed further on existing improbability. In the presence of a sufficient input of energy and in the absence of other limiting factors, the complexity of information-processing systems would have had to grow exponentially.

Conclusion

The concept that, as the universe evolves, its information content increases, is in opposition to the idea that the increase in entropy will, inevitably lead to the 'heat death' of the universe. However, it reflects the current rethink in physics as reviewed by Paul Davies in his book *The Cosmic Blueprint*: 'There exists alongside the entropy arrow another arrow of time, equally fundamental, and no less subtle in nature... the universe is progressing – through the steady growth of structure, organization and complexity – to ever more developed and elaborate states of matter and energy' (Davies, 1987, p. 20).

The fact that Figure 2.2 not only plots the relationship between information and entropy, but also allows one to represent the evolution of our universe, could be a mere coincidence. In addition, there is the uncertainty about the Big Bang theory itself. Nevertheless, even if it should turn out not to reflect anything but coincidence, the curve represents a useful metaphor. The increase in information in the universe would be as described above: first slow, then, as we cross the ordinate into the negative entropy/positive information quadrant, its exponential growth would become apparent. Complexity builds on preceding complexity. Increasing complexity provides increasing opportunities for further novelty and innovation. The greater the divergence, the greater also the possibility for creating higher levels of information processing as a result of synthesis – of combining two or more systems into a new whole. In this manner may we explain the emergence of information hierarchies – a matter to be explored next.

·3·

The Diversity of Information Processing

Introduction

The previous chapter should make it clear that in order to understand the evolution of the universe one must look to its base root: **the evolution of information-processing systems.** Looking at this evolution, it would come as no surprise that, after several thousand million years, many forms of information-processing systems (IPSs; the acronym was first used by Haefner (1988)) have emerged on this planet. This chapter will explore various IPSs to illustrate how advanced information systems have evolved different mechanisms to ***transmit*** information as messages, and to ***store*** such information in quite different and distinct forms.

The emergence of advanced IPSs would not be possible if physical information was not capable of being transformed from one form to another. That is, information, like energy, is continuously being transduced. This is particularly true in the human sphere of activity.

Information Transductions

The concept of ***transduction*** is well established in the transformation of one form of energy into another. For example, the steps leading from a power station to electric lighting in our home involve four distinct transductions: oil is burned in (1) a power generating plant to generate steam to drive (2) turbines, which coupled to (3) a generator, produce electricity which, in turn, causes (4) a lamp to emit light.

The first step converts ***chemical*** energy into ***heat***; the second involves changing the heat energy into ***mechanical energy***; the mechanical energy is transduced in the third step into ***electrical energy***, which, in turn, is transformed into ***electromagnetic radiation*** in the last step.

In all four of these transductions we are dealing with energy transducers designed and created by humans. However, long before the appearance of anything resembling humanity, highly sophisticated energy transducers had evolved in biosystems. The most important of these is the class of transducers capable of carrying out photosynthesis, which was probably first evolved by organisms related to the cyanobacteria and later by various organisms related to present-day algae. It is this energy transduction, in which sunshine is converted into chemical energy, which sustains virtually all forms of life (except for the chemosynthetic bacteria). The chemical energy created by this process is transduced by various forms of organism into almost all known forms of energy: mechanical energy for locomotion in animals, electricity in electric eels, light in fireflies, heat in warm blooded animals, and sound in social mammals, not to mention the huge amount of osmotic work carried out by virtually all cells, but in particular specialized cells and tissues such as nerve cells and kidneys.

Energy transductions are not limited to human or biological systems but occur in physical systems as well. The absorption of light by a black surface, causing it to heat up, or conversely, heating substances to a point where they emit light, exemplify physical energy transduction. Similarly, nuclear reactions, which occur in our Sun and other stars, convert nuclear energy into heat and light.

Just as there exist different forms of energy, readily interconvertible, so do there exist different forms of information. Although human information systems represent only one class of information, it is easiest to illustrate information transduction in human systems because human information itself may be stored and communicated in a wide variety of ways and be represented in many different forms.

Consider the information on this page. It is being transmitted to the reader's eye by light. The light striking the retina is converted to nerve impulses, which are propagated by the sequential depolarization of membranes. At the synapses between the nerve cells of the brain, the information is converted to pulses of chemical neurotransmitters which, in turn, trigger further neurological activities, branching in many directions. Ultimately, these events lead to a host of brain activities: short-term memory storage, comparison with existing information at many levels (from comparing pictures of the printed letters and words, and their meanings, to comparing the concepts of this article with the reader's view of the world), long-term memory storage and the myriad of other, still mysterious, thought processes associated with assimilating and analysing new information.

At some point in the future, the reader may wish to communicate the information to someone else by converting the patterns of neural information stored in the brain into sound waves via nerve impulses to the vocal chords. Sound waves represent a mechanical coding of information. The sound waves impinge upon the ears of the listener, where the information is now converted from pulses of mechanical energy into nerve impulses by the motion of microscopic hair-like organelles in the inner ear. These nerve impulses enter the brain of the listener, where the information will undergo a processing similar to that originally taking place in the brain of the reader.

Alternatively, the reader might speak into a telephone, where the information is transduced from patterns of compressed air molecules travelling at the speed of sound into electronic pulses travelling down a copper wire closer to the speed of light. These electronic pulses travelling down a wire might, in turn, be converted into pulses of light travelling down an optical fibre. Or the reader might speak into a microphone, either for broadcast, in which case the information is transduced into patterns of electromagnetic waves traversing the airwaves, or into a tape recorder, where the electronic pulses are converted into magnetic pulses, then 'frozen' into the tape by having atoms, responding to magnetism, arrange themselves physically into patterns of information within the tape.

The above represents one set of cycles which illustrate the communication of human information. Note what it entailed: the information was propagated as:

1. Patterns of light (from the book to the eye).
2. Pulses of membrane depolarization (from the eye to the brain).
3. Pulses of chemical substances (across the synapses between individual nerves).
4. Pulses of compressed air molecules, i.e. sound waves (emitted by the larynx of the speaker).
5. Pulses of mechanical distortion in liquid or solid (inside the ear or the telephone mouthpiece).
6. Pulses of electrons in a telephone wire.
7. Pulses of light in optical fibres.
8. Pulses of radio waves.
9. Pulses of magnetism (inside the ear piece of a telephone or the speaker of a radio).

The information was stored as printing in a book, in the human brain and on magnetic tape. The first involved patterns of dye molecules; the second, patterns of neural connections; and the last, patterns of magnetized regions. It was converted from one form into another in the retina of the eye, at the synapse between nerve cells, in the larynx, in the inner ear, in the telephone, in the radio transmitter, in the radio receiver and in the tape recorder.

The reader might also have decided to store the information on a computer, or in a file by photo-duplicating it, or by typing it out, thereby committing the information once more to patterns of dye molecules superimposed on paper molecules. From cave wall paintings and carvings on wood or stone, to bubble memories and satellite communication, human information is capable of being stored, transmitted and transduced in a very large number of ways – and the number continues to increase.

Note that the means of propagating information, as exemplified by the above list of nine, usually involved pulses of waves (light, sound, radio waves), pulses of electrons or pulses affecting matter or its organization. The fact that information may be divided up into small, discrete packets is utilized by communications engineers for packet switching to allow several users to use the same facility simultaneously. Although the communications engineers treat information as an abstract quantity, they do not follow this idea to its logical conclusion – that information exists in its own right. Perhaps part of the problem of recognizing and accepting the idea that information has physical reality and constitutes an intrinsic property of the universe stems from the fact that we ourselves are so deeply embedded in the processing, transducing and transmitting of it.

It took almost three centuries to establish the concept of energy and to calculate the various conversion factors. As far as information theory is concerned, we are, at best, somewhere in the 18th century equivalent of energy calculations. Consider one of the examples from above: reading out loud. How does the information content of the patterns of dye molecules on the printed page relate to the patterns of compressed air waves generated by the voice-box of the reader? The ***meaning*** may be the same – one might even argue, identical. However, the information content may be quite different, since the energy input required to create these patterns differs, and so does the input of information by devices whose previous formative history is so very different. That is, the ***printed*** word in a book has the individual history of the author, the editor, and various other people associated with the production and printing of the book. In addition, the energy input of putting patterns of dye molecules on sheets of paper involves not only the printing press, but the manufacture of dyes, paper and printing presses. In contrast, the ***spoken*** word has behind it not only the evolution of language (to be discussed in a later chapter), but the entire bio-evolution leading to the human species. The energy of the spoken word derives from exhaling air across the vocal apparatus; this is a far cry from that of an electrically operated printing press. Yet, symbolically, the printed word and its corresponding spoken word both have the same meaning.

These considerations lead us to explore the diversity of information-processing systems that have evolved in nature, in human culture and, more recently, in human technology.

The Human Social Information-Processing System

When most people think of information, they think of what in the present work is called ***human social*** information. For example, Brier's (1992, p. 74) rejection of information as a physical property of the universe is based on his perception that: '**information** basically is something **a person communicates to another person**' [bold text in original script]. Brier cites Machlup (1983) and Winograd and Flores (1987) as expressing similar views. In the present work, this anthropocentric conceptualization is wholly rejected. Such a limited definition of information would deny, for instance, that DNA carries information.

For this reason, we need to define ***human social*** information, and we need to define it both as ***human*** and as ***social***, for reasons indicted below. Human social information is information which is exchanged between human beings, either as individuals or as groups. It is an intensely human activity and it is the cardinal feature which distinguishes human society from other terrestrial animal societies. Its chief instrument of communication is the word, spoken or written, and until very recently such information was stored either as a form of oral tradition (including songs) or was recorded on tablets, in documents of various sorts or in books.

Human social communication does, of course, involve mechanisms other than human speech. At the interpersonal level our body language, derived from our primate heritage, can play a significant role. At the intergroup level, actions may speak louder than words. This is particularly true when two antagonistic groups work up to a conflict: a routine change in military deployment can easily precipitate a war as it may be interpreted by the other side as a signal of hostile intent. On a more positive side, global society has developed many channels of communication, from railways to financial markets, which not only transport words but other forms of information as well. That most powerful of present-day communicators – television – communicates much of its information in visual, rather than verbal, terms.

In so far as human social communication has been the subject of innumerable discussions, we need not go into the subject in greater detail at this time – we will return to the matter in a later chapter. Instead, our attention will be focused on animal communication: the exchange of information which takes place in non-human societies. Such exchanges occur not only in mammalian societies, such as those of monkeys and dolphins, where their sophistication has been well established, but also animals whose intelligence (by mammalian standards) is very low.

Ants: a Model for Social Information-Processing Systems

Compared with the human brain, an ant's brain has only of the order of one hundred thousandth of the number of neurons. Yet, this is sufficient to allow a typical ant to operate with somewhere between 10 and 20 different kinds of signal (Hölldobler and Wilson, 1990). Ants have no vocal organs and cannot talk. However, some species of ants have 'chirping' organs which are particularly useful if the ant is trapped underground and needs its nest mates to dig it out. Carpenter ants, as well as some other species, tap to produce sound vibrations which other ants are able to detect, while fire ants produce a thrumming sound by moving the abdomen up and down. As reported in *Science* (Holden, 1996), in the case of the fire ants, this thrumming produces several kinds of signal: a general alarm signal, an individual distress signal, an attack signal (to attack a caterpillar) and an 'all is well' signal.

In general, however, communication is carried out by three non-auditory mechanisms: (1) visual signals, such as body language, (2) tactile actions such as stroking and (3) by means of sets of sophisticated chemical signals. In so far as we are dealing with chemicals which send messages from one organism to another, these chemicals have been called ***semiochemicals***, and several categories of such substances have been defined:

1. ***Pheromones***: These are semiochemicals which are secreted by one individual with the intent of enhancing or changing the behaviour of another individual of the same species. The secretions usually

emanate from some gland and are tasted or smelled by the recipient.

2. ***Allomones***: These are semiochemicals which are used in communication across species; for example, a lure used by a predator to attract a prey species, or a noxious chemical released by a potential prey to thwart the predator. These chemicals are designed to favour the emitter but not the receiver.
3. ***Kairomones***: These are semiochemicals which are also used in communication across species, but with the reverse effect: they aid the receiver but not the emitter; for example, the odour trail left by a potential prey, which could lead a predator to it.

In addition to characterizing the semiochemicals in the above manner, they may also be divided into two broad classes based on their effects on the recipient. The first category results in a classic stimulus/response mechanism mediated entirely by the nervous system. The second acts on the endocrine and reproductive systems causing basic physiological changes in the recipient.

Both the range of the semiochemicals and the responses they elicit show an enormous variation throughout the 8800 ***known*** species of ants in this large insect family, the Formicidae.[1] Most of the pheromones are secreted by one of six key *exocrine* glands (in contrast to *endocrine* glands, which secrete into the bloodstream to achieve an internal effect). These glands, which secrete externally, serve a variety of functions in the diverse groups of ant species. Much of an ant's communication capability resides in these secretions.

The African weaver ant provides an illustration of the above. Hölldobler and Wilson list five distinct recruitment systems of this species, each of which involves a different set of signals. These five systems comprise (1) recruitment for food, (2) short-range recruitment to engage enemies, (3) long-range recruitment to engage enemies, (4) recruitment to new terrain and (5) emigration. All involve the use of chemical signals, most involve tactile signals, and some also involve visual signalling.

In each case, both the sending and the reception and interpretation of these various messages have presumably been hard-wired into the brains of these ants by evolution. This would account for the fact that the communication system is so rigid. The messages are so stereotyped that ants are readily fooled. For example, ant nests are frequently infested with insects and other pests which act as 'social parasites'. These parasites are

1 Hölldobler and Wilson (1990) estimate that there may exist over 20 000 species.

able to get ants to feed them (even sometimes while they devour the ants' larvae) simply by sending out a chemical signal, or, as in the case of a beetle, simply stroking an ant in a characteristic soliciting behaviour which mimics another ant begging for food. The beetle, by the way, in no way looks like an ant. Even more dramatic is the demonstration by Hölldobler that touching the appropriate parts of an ant's body with the tip of a human hair causes the ant to regurgitate food.

In spite of the rigidity of the social communication system of ants, the system must be regarded as a formal information processing system. Even if the ant system does not exhibit the same range and flexibility as the human social system, and even if the nature of the messages is different, the two of them follow the same rule of social communication: the sender sends a message to a receiver who then acts upon the message. If the receiver responds to the message in a manner which satisfies the sender, the feedback loop is completed and the sender stops sending its message. Note that in all this we did not invoke 'understanding' as a criterion for an information processing system. ***Understanding*** of a message, or its ***meaning***, is a different issue and requires an analysis of the evolution of intelligence – a matter discussed by the author previously (Stonier, 1992) and a matter to which we will return in later chapters.

The Neurological Information Processing System

Although the brain might be considered as a society of neurons, it is an IPS which works on an entirely different set of principles. Whereas in human and animal societies the information flowing between two individuals as a message usually results in instantaneous feedback – i.e. that the message has been understood (or not) – a message flowing along nerve cells is largely a one-way phenomenon, although there is much feedback and resonance in the brain itself (see the later chapters). Each neuron seems to act merely as a conduit of information. But, as we know, the matter is much more complex than that. For one thing, the neuron may remember that it has been such a conduit. It may strengthen the synapses which have been used in the impulse conduction, leading to what is called ***long-term potentiation***. Such long-term potentiation along certain pathways could lead to resonances, which, in turn, provide feedback. Thus, although there may not be an immediate feedback between interacting neurons, feedback mechanisms do abound in the overall organization of the nervous system, and in particular in the brain.

There is another unique feature which characterizes the neurological IPS: the system not only conducts and processes information, it also acts as its own memory. This is very different from other IPSs, such as a computer, a cell (whose memory is stored in its DNA) or a human being (whose memory is stored in its brain).

Likewise, the transport of information is very different from other systems. Information is transmitted by means of the progressive depolarization of the outer membrane along a single nerve fibre, and by the transmission of pulses of various chemicals (neurotransmitters) across the synapses which connect the neurons.

To sum up, the physical basis of transporting messages in the brain involves patterns of neural pathways, and, interestingly enough, by strengthening the connections across synapses (or allowing them to decay), patterns of neural connections also provide the means for storing information. It must be clear, even at this stage of our ignorance, that ***neurological*** information-processing systems operate on principles and utilize tools entirely different from those used either by social or electronic information-processing systems.

The Immune System

Some theorists consider the immune system as a form of linguistic system: '...the immune system embodies a degree of complexity which suggests some more or less superficial though striking analogies with human language' (Jerne, quoted by Tauber (1944)). The immune system has the capacity to produce ten million different proteins (over a thousand-fold more than all other body proteins taken together). Each protein is made up of hundreds of amino acids. Jerne considers the former as a sentence and the latter as individual words.

The foreign antigens, too, may be regarded as sentences. Thus, the immunoglobulins produced by the body and the foreign proteins are able to communicate because their respective word orders possess a kind of symmetry – a symmetry which leads to a process of information exchange. Of all the myriad of antibody–antigen interactions which could, in theory, take place, it is only this specific antibody and this specific antigen which 'speak to each other' and coalesce.[2]

---

2 This is not entirely true: the same idiotope may be shared by antibodies with various specificities, and antibodies can, on occasion, bind more than one antigen.

For the immune system as a whole, the system uses the information provided by the antigen to clone a large number of antibodies. Thus, the immune system is an information processing system with a multiplicity of feedback loops. An invading antigen is perceived and responded to by the immune system on the basis of its molecular structure – as it were, its protein 'sentences'. The antigen therefore constitutes a message which, by the specificity of its response, has meaning for the immune system.

Maintaining Body Fluid Balance[3]

A biological subsystem which processes information to achieve homoeostasis, thereby maintaining the stability of the system, is exemplified by the complex biological and behavioural feedback loops which control the amount of salt and water in our body fluids (Stricker and Verbalis, 1988). Normally, thirst appears to be controlled by an osmoregulatory system which involves a simple, single negative feedback loop. The brain contains certain sensors called 'osmoreceptors'. When these become dehydrated, they activate a central system which induces the sensation of thirst. This results in a behavioural response: we drink fluids until the dehydration is corrected, and the goal of the system – to maintain a reasonably constant volume of body fluids – has been attained.

Suppose fluid intake has been excessive. The surplus water is then excreted in the urine. Excessive drinking causes the concentration of solutes in the extra-cellular fluids of the body to fall below the allowed tolerance of 1–2%. This leads to an *inhibiting* of the secretion from the pituitary gland vasopressin, the antidiuretic hormone which promotes water conservation in the kidneys. Thus the kidneys release water. At the same time, thirst is also further inhibited.

If we eat salty food, such that the concentration of sodium in the body's fluid exceeds a threshold, the body interprets this as a deficiency of water because the brain's osmoreceptors once more signal that the body is in a relatively dehydrated state. The loop is the same as described in the first paragraph above.

What happens when we are in an accident in which we lose blood? Or to simplify matters, what happens when we donate a pint of blood to a blood bank? The body is short of a pint of fluid. The first priority is to stabilize the blood pressure. There is a neural input into the brain from the stretch receptors embedded in the smooth muscle wall at various sites of

3 This section has appeared previously in Stonier (1992).

the circulatory system. These 'baroreceptors' send signals which cause the pituitary to release vasopressin. Vasopressin, as discussed above, is an antidiuretic which causes the kidney to retain water. It also acts as a vasoconstrictor, causing the blood vessels to constrict and thereby increasing the blood pressure. Finally, the same signals appear to stimulate thirst as well.

Drinking plain water, however, is a poor substitute for the loss of blood. The lost blood also contained sodium. Experiments with rats have shown that the thirst is abated well before the quantity of lost fluid has been replaced. However, rats, given access to a concentrated salt solution as well as plain water will consume a mix of the two, so as to both restore the lost body fluids and maintain the right osmotic pressure. The stimulus for sodium appetite appears to involve a number of other hormones – renin, angiotensin, aldosterone, oxytocin – acting at times in combination to stimulate positive behaviour or, alternatively, counteract the inhibitory effect of hormones or nerve centres of the brain. Obviously, these involve complex feedback loops because the body is now attempting to reset the system with respect to two sets of parameters: the first is to replace the lost body fluid and restore normal blood pressure, and the second is to retain normal osmotic pressure by ensuring that the water intake is balanced by an appropriate sodium intake.

Other Bodily Subsystems

Higher animals contain many examples of such self-regulating subsystems, which not only *detect* altered organizational states but then *process* the information and *act* on that information to return the system to its original state. Notice how in the above example of complex feedback loops, not only are certain organic chemicals (hormones) acting as messengers, but a message may also consist of an inorganic chemical, NaCl, and even a non-material physical state: blood pressure.

Before we leave the subject, it might be well to remind ourselves why these information-processing systems have tended to be overlooked. The psychologist Earnest Rossi (1993, p. 158) stated it succinctly when he wrote that: 'traditional textbooks of medicine... are usually organized around anatomical and physiological functions in such a way that the signaling and communication processes... are usually obscured, lost, or just thrown in as a sort of metaphor afterthought about how things work at a molecular level'. Rossi himself recognizes four major self-regulating systems which are responsible for the normal functioning of the body.

All of these use chemical messengers to communicate, but the nature of the substances is different. His four categories are:

- the autonomic nervous system utilizing neurotransmitters
- the endocrine system using hormones
- the immune system, which uses cytokines
- the neuropeptide system, which relies on neuropeptides

To some extent the cytokines and neuropeptides could be considered as forms of hormones not secreted by special organs (the endocrine glands).

It is not the function of the present work to review this material further but to drive home the point that the human body is an information processing system of extreme complexity which is made up of numerous information-processing ***sub***systems, that is, IPSs which are one hierarchical level below that of the body itself. Let us now go down the IPS hierarchy scale to a further, lower level.

Cellular Information-Processing Systems

Cells contain numerous information-processing systems. In the light of the previous discussion, envisage a cell responding to a hormone. The hormone is a chemical message that causes the cell to engage in some sort of metabolic response. The message is encoded in the arrangement of the atoms comprising the hormone molecule – to be precise, the information is transcribed as a ***spatial pattern of atoms***. This pattern is recognized by certain receptor molecules on the surface of the target cell – molecules which then, themselves, act as messengers carrying the information across the cell membrane to the interior of the cell.

Spatial patterns of atoms are the key to all information processing carried out by individual cells. As with the immune system, it is helpful to use an analogy with the human language system, in which the individual subunits – amino acids, nucleotides, sugars and fatty acids – are considered as words, while their respective polymers – protein, nucleic acids, polysaccharides and lipids – are considered as sentences.

This concept is most clearly demonstrated by ***messenger-RNA***. Messenger-RNA is made up of a long string of nucleotides, although only four kinds are involved: adenine, cytosine, guanine and uracil. These may be considered as 'letters' which make up 'words'. However, in the language of DNA and RNA there is a strict syntactical rule: all words are allowed to contain no more and no fewer than three such letters. If this severely limits the kinds of words possible, there is no such limitation on the number of words that can be strung together. Very long sentences may be created involving thousands of such words. (With suitable preparation, the messenger-RNA may actually be seen as long, thin strands in an electron microscope.) For each messenger-RNA, the ***sequence*** of words is created by the DNA in the nucleus: the DNA provides a template from which the RNA makes a copy.

True to its name, once a molecule of messenger-RNA has been created, it moves from the nucleus into the cytoplasm of the cell where tiny organelles called ***ribosomes*** 'read' the information on the messenger-RNA as if on a tickertape. The words on the RNA are read in groups of three, where each triplet designates a specific amino acid. The amino acids are assembled in accordance with the information provided by the RNA, thereby creating a specific protein.

The flow of information from DNA to RNA to protein is a classic case of information transduction. Its human metaphor would be the construction of a skyscraper: the architect's sketches become transduced into precise engineering plans which in turn become transduced into an organized physical structure. The sketches are not the same as the blueprints, and the blueprints are not the same as the building. Yet the initial information is there in all three forms, although at each step, as the information is processed, the information content is increased.

The proteins, as end products, could be considered analogous to the building. Some proteins are, indeed, structural proteins which when combined with membranes and other substances (cellulose in plants, silicon in diatoms etc.) give a cell its shape and physical structure. Most proteins, however, have other functions, such as enzyme catalysis, transport protein, activators (kinases) etc. The numerous feedback loops which govern all of these activities represent other sets of information-processing systems and subsystems, described from an information point of view by Pedro Marijuán (1993, 1996). Marijuán recognizes three different degrees of complexity which appear as collective properties among the network of enzymes: (1) self-organization, (2) self-reshaping and (3) self-modification. He further considers that the enzyme network, encapsulated in a DNA world, constitutes a genuine computational core, a ***computational society***. Suffice it here to state that this complexity is a product of the information residing in the ***patterns of nucleotides*** making up a cell's DNA. It is this repository of information that represents the cell's most vital store of information: the information needed to reproduce itself.

In order for a cell to reproduce itself, it must first live, i.e. ingest nutrients, convert them to living matter, excrete waste products, grow and finally enter a cell division cycle. As one goes from a primitive bacterial cell to a more complex 'eukaryotic' cell, such as yeast, the DNA text governing the process of living becomes more extensive and more elaborate. In a bacterium, the number of nucleotides total several hundred thousand, organized into approximately 2000 genes. In yeast, the 12.5 million nucleotides are organized into 6000 genes. Multicellular organisms, whose sophisticated organization took much longer to evolve, are one or two orders of magnitude greater than that. The human genome project has identified of the order of 2.5 billion nucleotides (Bussey, 1996).

Mechanical Information-Processing Systems

All of the above examples involved human or biological information systems. Let us turn now to three mechanical IPSs to illustrate how different such systems may be. The first is an early example of automation which was devised during the early part of the industrial revolution: the ***governor*** of a steam engine. Note that the name given to this device, 'governor', could be considered as semantic evidence of an intuitive perception by the 18th century engineers that the governor was no ordinary bit of machinery.

This device automatically regulates the speed of, for example, a steam locomotive. There are several types, but the best known is the flyball governor. In this device, two 'arms' hang down from the top of a vertical shaft. Attached to each is a heavy ball. These arms spin around the vertical shaft. The faster they spin the farther out and up they move because of the centrifugal force exercised on the balls by the rotation. The arms may end up fully outstretched horizontally if the shaft rotates fast enough. As they move further out and up, they lift up a ring around the shaft which is connected to a throttle. The faster the shaft turns – the shaft being driven by the steam engine itself – the greater the force on the spinning balls and therefore the greater the lift of the upward ring. As the ring is being lifted upward, it causes the engine to throttle down.

The governor on a steam engine comprises one of the simplest forms of negative feedback. The faster the steam engine works the faster the governor rotates. The faster the governor rotates, the more it throttles the steam engine down. As the steam engine slows up, the balls of the governor rotate more slowly and the arms drop. When the arms drop, the throttle opens up again. When the throttle opens up, the steam engine starts working faster again... until the speed of the governor's balls causes a throttling down once more.

If the above is confusing to a reader not acquainted with such a device, the point is that the governor is fed information from the steam engine as to how fast it is going, and depending how it has been set (by its human controller), it will respond by sending a message back to the engine to slow down or speed up. It is perhaps one of the simplest of all information-processing systems possible. It reflects the fact that it emerged during the early evolution of humanly created power machinery.

Most readers will be acquainted with a more recent and more commonplace example of this kind of feedback mechanism: the ***thermostat*** in people's houses. When the temperature rises in a house as a result of heat being supplied by a central heating system, the thermostat detects this and the expanding metals cause a switch to be forced open. When the electrical switch is forced open no further fuel is supplied to the boiler. The boiler shuts down. As the temperature in the house falls, the metal detector of the thermostat cools, and in due course, closes the switch once again. Once the switch is closed, the boiler is turned on again and the temperature will rise once more as the house is being heated.

Here again, we are dealing with a simple feedback loop. The house furnace sends a message to the thermostat by heating its ambient air. The thermostat responds by sending a message back via an electrical circuit.

A third example, this time an electro-mechanical system, is illustrated by what happens in an electrical power grid (Wiener, 1961). In the early days of alternating current engineering, attempts were made to connect generators of the same constant voltage type used in modern generating systems in 'series' rather than in 'parallel'. It was found that the interaction of the individual generators in frequency was a repulsion rather than an attraction. The result was that the system was impossibly unstable unless the rotating parts of the individual generators were connected rigidly by a common shaft or gearing. On the other hand the 'busbar connection' of generators, hooking them up in parallel, proved to have an intrinsic stability which made it possible to unite generators at different stations into a single self-containing system. Wiener uses a biological analogy: the ***parallel*** system had a better homoeostasis than the series system and therefore survived; in contrast, the ***series*** system eliminated itself by natural selection.

From our point of view, the individual electric generators which make up a power grid could be considered as a 'society' in which the individual generators are able to send signals to each other to keep the frequency of their output in phase. As such, the system achieves a ***resonance*** which binds the individual units together into a coherent whole. This is an important principle which allows simpler systems to evolve into more complex, differentiated ones. In fact, it should be axiomatic that one cannot understand the physical basis of IPSs, nor their evolution, if one does not understand the concept of resonance and its significance in physical processes.

Physical Information-Processing Systems

The phenomenon of electronic resonance plays an enormously important role in molecular and atomic interactions. It is the structural properties of both the arrangement of atoms within molecules and the atomic architecture itself which give each interaction its specificity. By definition, these structural organizational properties are all a reflection of the information content of the system.

A good example is the benzene molecule. This molecule is made up of six carbon atoms arranged into a hexagon-like ring structure. To each carbon atom is linked one hydrogen atom. The six carbon atoms themselves are linked to each other by three single and three double bonds. These bonds alternate continuously, so that one carbon atom may have one double bond to the left and a single bond to the right, while in the next moment the situation may be reversed. The shared electrons which are responsible for these bonds are part of a resonance system referred to as the 'pi electrons'. In a sense, these pi electrons communicate between the 'society of six carbon atoms' to stabilize the entire benzene structure. The resonance achieved by the benzene molecule acts as a kind of homoeostatic mechanism, which, for an organic compound, makes benzene an extremely stable molecule.

All bits of matter, from sub-atomic particles to the most elaborate of molecules, are able to maintain their integrity by having an exchange of information among the constituents which make up such matter. This exchange of information takes place through the various physical forces which hold together the constituent members. In probably all cases this involves one or more resonating systems. Resonating systems may be viewed as finely attuned information feedback systems.

Although it is the physical properties of resonating systems which cause such systems to resonate, these properties are really dependent on information inputs which determine the spatial and temporal parameters of the system. It is the size of the sound box and the length of the string plucked which determine the note a violin plays, and the sizes of resistors, capacitors and induction coils which determine the frequency at which a transmitter emits radio waves. It is by varying not the energy inputs but the information inputs that determines the wavelength of the output, whether sound (violin) or electromagnetic radiation (transmitter). Resonating systems may be viewed as a form of physical IPS.

Electronic Information-Processing Systems

More recently, the technical evolution of modern societies has created an entirely new type of information processing system: ***computers***. Here information is transmitted electronically along copper wires while the information is stored as patterns of on/off switches. In this case, we are dealing with an ***electronic*** information-processing system – a system fundamentally different from all the other information-processing systems discussed above. Consider, for example, the human social IPS: when two people talk to each other, the ***type*** of information processing is very different from when two computers communicate with each other. In the first instance, as we have discussed above, the information is encoded as spoken words carried to the listener as patterns of compressed air waves emanating from the speaker's voice box. In the latter instance, the information is transmitted as patterns of pulses of electrons (actually signifying only two words: 'on' and 'off'). In like manner, storing the transmitted information as written words is entirely different from storing it in the form of patterns of solid state on/off switches. We will discuss computers in greater detail in a later chapter.

General Discussion

What must strike one from the above description of the various IPSs is the diversity of methods which have evolved to communicate between various parts of a system, the nature in which the information may be processed, and, if it is a sufficiently advanced system, the method of storage. What must also emerge is that information may exist in many forms, yet it is also the raw material which underlies all of these processes.

Starting with one of the most primitive of all information systems – the house thermostat – it must become clear that it is a nonsense to analyse that system by means of matter and energy alone. The thermostat supplies neither raw material nor energy (except indirectly by closing an electrical switch which activates a fossil fuel supply). In fact, it may actually ***withhold*** energy when its sensor determines that the furnace ought to shut down. The main function of the thermostat is purely an informational one.

The thermostat also provides an example of a purely physical system in which the feedback relies on two very different message systems. The one sent by the thermostat involves an electric current to start up the motor, which pumps fuel to the furnace. The one sent by the furnace is hot air. That a physical state may act as a message is not so strange if we recall that blood pressure is a message in maintaining a body's homoeostasis. Another example, not discussed before, is that both light and gravity act on certain receptors of plant cells, orienting the shoots and the roots.

The thermostat, as well as other primitive mechanical devices such as the governor, provides us with another insight: the evolutionary beginnings of a new line of IPSs. Under the impact of the Industrial Revolution a whole new branch of information-processing systems began to emerge. We have examined, if briefly, three of these mechanical IPSs – examples chosen because they all showed feedback loops. This should not be construed as being a prerequisite for all information-processing systems, since the flow of information may be one-way through an 'open system'. What is important is that with the Industrial Revolution something totally new emerged. The same may be said of the Electronic Revolution, which differed from the Industrial Revolution in that the latter extended the human musculature, while the former extended its nervous system (Stonier, 1983). The ultimate achievement of the Electronic Revolution was the computer, an extension of the human brain.

The nature of the information inside people's heads must be different from that contained inside a computer. When compared with a computer, the human brain exhibits substantially greater complexity in at least three areas (Stonier, 1984; see also Appendix D). First, the circuitry is incredibly more complex: not only does the brain contain of the order of 10^{11} cells, but a single brain cell may be connected to many thousands of other cells. Each neuron, in turn, may prove to be equivalent to a transputer rather than a transistor. Second, the transmission system is different. In a computer one is dealing with electrons moving along a conductor. Nerve impulses, in contrast, involve the progressive depolarization of membranes coupled to the transmission of the information across the synapses by means of chemical signals. This device allows the transmission system to be regulated much more delicately and relates to the third major difference: information handling in the present generation of computers is digital. In the human nervous system there exist dozens of neurotransmitters and other related substances which can enhance or inhibit nerve impulses – the whole system is a finely tuned and integrated network of analogue devices.

The point emphasized in this chapter is that the transmission of information via a copper wire (or optical fibre) is entirely different from the depolarization of membranes along a nerve fibre and the transport of chemicals across a synapse. Likewise, the storage of information as patterns of on/off switches in a computer's memory is a far cry from the pattern of strengthened neural pathways in the brain.

This leads to the concept of the ***physical basis of information transfer (the message)*** – that is, the raw material from which a message is constructed – which in human social information processing involves coding the morpheme (in the case of the spoken word) and the letter (in the case of written, phonetic language). The message results from having created patterns of morphemes or letters. This is very different from coding on/off switches and producing messages by stringing together pulses of electrons, which is the basis of electronic IPSs.

The evolution of biological information systems began with the origin of life. During subsequent evolution, a variety of information systems have emerged and, in turn, evolved to contribute to the further evolution of the living systems which spawned them (Conrad, 1996). Except for the fact that we are always dealing with patterns – an inevitable consequence of the presence of information – there are no similarities in the physical bases of the messages between cellular information-processing systems and the others mentioned above (although the synaptic transmission of the neurological information processing system relies – and must have evolved from – the cellular mode of information transmission).

Each of the above systems may be broken up into numerous further subsystems. Human social information-processing systems may be broken up into spoken and written language systems. One could consider each spoken language as a separate subsystem, each with its own syntax. There are other human systems which rely more on pictures: these range from cave paintings and medieval church windows to pictographic written languages and modern traffic signs. In human social information-processing systems, body language, including facial expressions (such as a smile), also represents an important message system.

The opposite also holds. Different types of information-processing system may merge to create new kinds of systems. The spoken word may be transmitted electronically as pulses down a telephone line. A written document may, likewise, be sent as a fax or as email, the latter providing an example of a mix of a social and an electronic information-processing system. Another recent such mix is reflected in the advent of neurocomputers. These represent the marriage of electronic and neurological information-processing systems.

The merging of existing systems is one of the major mechanisms by which information systems evolve. It is one way of creating hierarchies of complexity, a problem which has also been occupying Klaus Haefner at the University of Bremen. Haefner (1991) posits the idea that the entire universe is made up of information-processing systems – ranging from the smallest physical particles to human societies and technological systems – and that all such systems evolve.

Conclusion

Our universe is filled with many different kinds of information-processing systems (IPSs). Different IPSs operate on entirely different principles, both in how they transmit information and in how they store and retrieve it. Some are closed systems with feedback loops, others are open systems. What should become apparent is that information processing is going on all around us and within us. It organizes the universe, it sustains life, it gives shape to culture, and it is the basis of our thoughts.

· 4 ·

The Spectrum of Intelligence

Introduction

In order for our discussion to proceed to the phenomenon of ***meaning*** we must first digress and examine the phenomenon of ***intelligence***. It is hoped that the previous chapters have established in the mind of the reader that information is a basic, structural property of the universe and that information-processing systems are ubiquitous. Information-processing systems may manifest the property of intelligence, a property which is a function of the sophistication of the information-processing system. In general, the higher the level of intelligence, the more evolved the system and the greater its sophistication.

The natural history of intelligence – from its murky origin as proto-intelligence to its future manifestation as humanity evolves beyond itself – has been covered in detail by an earlier work, *Beyond Information* (Stonier, 1992). This and the next three chapters are a summary and extension of this work, which, with the present earlier chapters, will lay the groundwork for our further exploration of the phenomenon of meaning.

Intelligence: the Basic Concept

A pocket calculator can extract the square root of a number in a twinkling of an eye – much faster than we can. Does that mean that a pocket calculator is more intelligent than a human being? What does it mean to be intelligent?

For the purposes of the present discussion, just as *information* is a basic, physical property of the universe, so is *intelligence* considered to be a property of advanced information-processing systems. *Intelligence* is a product of the evolution of information-processing systems (IPSs).

Intelligence is a state or a condition. As a property of advanced IPSs, it can be ascertained only by observing the dynamics or behaviour of the system. That is, 'intelligence' can only be ascertained in terms of 'intelligent behaviour'. For example, it is impossible to assess the true intelligence of a child when it is asleep. We assess the intelligence of a child by observing how it responds to its environment (including the questions posed by a psychologist).

Intelligence is an irrelevance to a system in a static state. On the other hand, a dynamic system in a changing environment may be able to remain stable only by virtue of engaging in intelligent behaviour. ***Intelligence*** is required to retain a system's integrity by adjusting it to the environmental changes.

Intelligent activity, for the most part, involves a system's ability to analyse its environment and then to make an intelligent response. An intelligent response may result in one of three states:

- The system has enhanced its own *survivability*.
- The system has enhanced its own *reproducibility*.
- If the system is goal-oriented, it has enhanced the *achievement* of that goal.

The above broadens the concept of intelligence as it is generally used in everyday life, while at the same time attempting to define 'intelligence' in fairly precise terms. However, because one is dealing with a phenomenon which has evolved, its origins prevent us from drawing too fine a line between the intelligent and the non-intelligent: there exist phenomena which are borderline. These we will consider to involve sub-intelligent, or *proto-intelligent* behaviour.

Finally, all intelligent systems engage in ***information processing***. They also exhibit the capacity to learn. ***Learning***, in the present discussion, is defined as the ability to modify intelligent activity on the basis of previous experience (real or vicarious).

Intelligence: a Spectrum of Phenomena

Intuitively, even the most anthropocentric of us feels that higher animals possess some form of intelligence. No dog lover can really believe that dogs are devoid of intelligence. Even the owners of pet fish recognize personalities in the various fish in their tank – some are more aggressive, others placid; some are keen to school, and some are solitary and defend their territories vigorously – and all learn quickly the signals that they are about to be fed. It is in this ability to learn, whether by fish or fowl, dog or human, that we recognize the glimmering of intelligence.

It is true that fish are not 'intelligent' the way we are. We are, after all, by far the most intelligent creatures on Earth – at least so we are led to believe. However, if we concede some sort of intelligence to animals – fish, parrots, dogs and monkeys – then we recognize that intelligence may exist as a spectrum of phenomena – that it is not an all-or-nothing affair. It is this intuition that becomes crucial for developing an understanding of the phenomenon of intelligence. It is also this intuition which was formalized by that most profound of observers of biological phenomena – Charles Darwin. Darwin proposed the doctrine of 'mental continuity': the differences in mental activity between various species of animals are differences of degree, not of kind (Menzel, 1986).

If intelligence must be viewed as a *spectrum* of phenomena, such a spectrum is not a simple, single-dimensional one. Mammalian 'intelligence' – the ability to perceive and analyse an organism's environment – is heavily dependent on eyes in order to see and interpret various aspects of the environment (blind species, such as moles, are obviously excepted). One could consider, therefore, our (human) perception of colour to be more sophisticated and superior to that of an animal, such as a cat, which is red-blind. We, like other higher primates, have no difficulty in picking out red cherries on a green cherry tree. Cats do: cats would see the leaves and fruit as various shades of grey. So obviously human vision is superior. Or is it? Cats are not particularly interested in cherries. Instead cats' eyes are designed for night vision and allow cats to move rapidly and stealthily in the dark, where we would stumble around. So which visual system is superior?

The visual perception of objects, movement and changes in our environment is only a part of our sensory system which helps us to analyse and respond intelligently to our surroundings. Cognitive scientists, in the past, have tried to measure the 'intelligence quotient' (IQ) or some other single parameter of intelligence in human individuals, or across species. Their results have been given a weight of authority wholly unjustified. The manifestations of human individual intelligence are determined by a variety of cultural and environmental forces, as well as genetic ones: a multidimensional input. Even 'identical' twins will have differences in their brain structures at birth (Edelman, reviewed by Rosenfield (1988, p. 186)). Measurements across species become increasingly tenuous. By the time we deal with brains – even insect brains – we are dealing with advanced forms of complexity. We should compare intelligent behaviour with other physical traits, such as size, strength or longevity. In mammals, both mice and elephants are well adapted to their environments, yet vary greatly in these three traits. It is the evolutionary history and the ecological requirements which dictate the size, strength and longevity of organisms. The same may be said of intelligence.

Therefore, for any given organism, the nature and sophistication of its intelligence is determined by its environment. If intelligence comprises a spectrum of phenomena, it is a complex, multifaceted one indeed. It is not some easily measured quantity which expresses itself merely as the ability to reason logically. Very few of humanity's great insights were derived as a result of logical deductions (only their veracity was justified, after the fact, by the application of reasoning).

Beyond Information: the Evolution of Intelligence

The idea that intelligence involves a *broad* spectrum of phenomena requires an expansion of what is usually thought of as 'intelligence'. The evolution of information-processing systems from non-intelligent to proto-intelligent and, finally, to systems clearly exhibiting intelligence, involves fuzzy demarcations with substantial overlaps. We are forced to make almost arbitrary decisions when we attempt to classify intelligent systems. Certainly, in the light of our limited knowledge, any taxonomy of intelligence proposed at this time will be open to debate at some point in the future. Probably, in the long run, it will not matter so much where, exactly, the lines are drawn, as that it becomes clear that one is dealing with a constellation of phenomena which have evolved from relatively simple information-processing systems to the highly complex – from fundamental physical particles to a human global collective intelligence. Let us retrace the evolution of intelligence by going backwards in time towards increasingly simpler forms.

Human Intelligence

The common notion of 'intelligence' tends to focus almost entirely on the individual human being. Thus we administer IQ tests to individuals. We would never dream of measuring the IQ of a society. Yet human intelligence is a manifest property exhibited not only by an individual, but also by a group of people, by entire countries, or even by global society as a whole. One of the major differences between the human species and other organisms is the fact that humans were able to develop an effective collective memory. This was done through developing an oral and subsequently a written cultural tradition which passed information from one generation to the next. By being able to store information and retrieve it across both time and space, the human species was able to develop a collective store of information. This process, in due course, became institutionalized. In Western Europe, for example, monasteries, churches, archives, museums, universities and libraries became a part of, and reinforced, cultural traditions. Most of these technologies also developed in other advanced civilizations around the world.

It has always been a mystery as to why, technologically speaking, Europe began to outstrip all other cultures during the 16th and 17th centuries. This is particularly puzzling *vis-à-vis* China, whose technological superiority was unquestioned right up to and through Europe's Middle Ages. The puzzle becomes resolved if one invokes the concept of ***collective intelligence***. As discussed in an earlier work (Stonier, 1992, pp. 89–98), the introduction from the Far East of printing technology which could be coupled to a phonetic alphabet (rather than the cumbersome pictographic one of China) led to a phase transition in the European culture. Using Latin as a trans-European language, a relatively literate Europe (compared with other cultures where only scholars and scribes could read) became a hot-bed of new ideas as printed works began to create a reliable collective memory. No persuasive argument can be made that individual Europeans had become smarter or dumber than individual Chinese. But the number and diversity of printed tracts traversing European society demonstrated the new creative power of the European collective intelligence.

To emphasize further the role of the printing press in the flowering of European intellectual and technical life: could a ***Royal Society*** publish its ***Proceedings*** if it had to rely on the older method of having monks transcribing it? Could such a ***Society*** be founded if printing and publishing were not already well established?

As the number of published works grew, libraries grew both in size and in number. Today, we are witnessing, once more, an increase in the collective intelligence, but this time of ***global*** society. The newly developed ability to store and retrieve information, as well as to transmit it by electronic means, not only allows more information to be stored per unit space, but more importantly the computer permits much faster information storage, retrieval and dissemination.

This is a major improvement in our collective memory: we have entered an era which will see all major printed databases – library catalogues, the entire content of encyclopaedias, dictionaries, government statistics, telephone directories, professional society memberships, train timetables, airline schedules, stock market quotations, etc. – become available to all homes via the World Wide Web or other networks, 24 hours a day, with a markedly increased ease of access.

In general, a ***group*** is much more capable of solving problems – and therefore in engaging in intelligent behaviour – than the average individual within such a group. There may be exceptions to this, but suffice it here to state that if intelligence comprises a spectrum, human ***collective intelligence*** represents the high end of the spectrum of known phenomena.

Whether ***individual human*** intelligence – i.e. an individual human being, isolated educationally from all human culture, but stimulated and nourished sufficiently to attain normal functioning – is a match for the collective intelligence of a group of wild dogs, or a troop of baboons, remains a matter of speculation. In fact, at present, so does the question of the relative individual, native intelligence of humans vs. certain species of dolphins, and possibly even other animals with larger brains than humans, such as elephants. However, individual human beings raised in a normal human environment, absorbing information from their cultural surroundings, will be considered as next in line in the spectrum of intelligent phenomena. That is, in the spectrum of known intelligences, ***individual*** human intelligence ranks below human ***collective*** intelligence and above animal intelligence.

Sub-Human Intelligence

Most people intuitively believe higher animals to be intelligent. Undoubtedly this reflects the fact that higher animals are able to engage in one form of higher intelligent activity – learning. Trained dogs, circus seals, street monkeys and aquarium dolphins all demonstrate that certain animals possess the capacity to learn complex routines. This intuition, however, is not shared by all. The author has held numerous discussions with a wide range of scientists and related professionals. The majority will draw the line in one of two places. The first group accepts that animals capable of learning exhibit some form of intelligence; the second, however, considers a creature to be intelligent only if it is able to interpret what it has learned. The ability to interpret presupposes that the animal has some model of the world in its head which becomes altered.

This view, at first glance, would appear to provide a much more rigorous, if limiting, definition of intelligence: it would seem to exclude all animals with the possible exception of higher primates and perhaps certain dolphins. But does it? A bee can be trained to find a dish of sugar solution at a specific location. It can be retrained to find it at another location. It can communicate that information to other bees so that they fly unerringly to the new find. How could this be possible if bees did not have a model of the world in their brains?

Bees must not only possess some sort of model of the world in their brains, but be able to modify that model on the basis of semiotic activity – utilizing abstract visual and auditory symbols to communicate. Surely, by the above criterion, a bee exhibits intelligence.

The brain of a bee is about the size of a pinhead. Its intelligence cannot be as advanced as that of mammals. Yet it engages in intelligent behaviour. Once more, we are forced to conclude that intelligence involves a spectrum of phenomena.

Our closest taxonomic relatives are the chimpanzees. Their intelligence is not only much greater than that of a bee, it begins to approach our own. One feature of advanced animal intelligence is the ability to use tools in a goal-oriented fashion – a behaviour observed in many animals – and, more advanced still, to *create* tools. It was Jane Goodall (1971) who first established that chimpanzees in the wild not only used tools, but actually *fashioned* them. Her reports, and others, indicate that these apes use twigs and rods to 'fish' for termites in termite mounds or obtain honey from beehives; sticks to probe trees for possible insects; sponges made from chewed leaves to sop up rainwater in tree hollows not directly accessible to the mouth or to wipe remnants of brain from inside the skull of a killed baboon; leaves as toilet paper; stout sticks as levers to pry open banana boxes or enlarge the opening of an underground bee's nest; rocks as hammers; and rocks, sticks, serrated leaves and branches as weapons. It was Goodall who observed that both the fishing sticks and the leaf sponges involved modifying natural objects to use as tools. Primitive as such tool-making may be, it must involve a chain of problem-solving reasoning to achieve a desired goal.

If tool-making represents one form of intelligent behaviour which involves deep insight into the physical environment, the observations of primatologists such as Kummer (1982) have demonstrated that monkeys are capable of stealth and deception – a form of behaviour which can occur only if animals possess a high degree of insight into the social situation at hand. Two examples will suffice:

- A young female hamadryas baboon in oestrus repeatedly mated with a young male behind a rock where her adult male troop leader could not see her. Between matings she moved to peek at the leader, or even, on occasion, went to him and presented herself before returning to her hiding place to continue intercourse with her young mate.

- In an experiment to test the 'loyalty' of female gelada baboons, individual females were confined with individual males who were not their mates. Their male mates were kept on the other side of a concrete wall where they could hear but could not see what was going on. When females began to groom the illicit males, instead of the loud vocalization which normally accompanies such activity the latter would restrain themselves and remain quiet.

Another set of observations which attest to the social intelligence of non-human primates is described by Menzel (1975): in reviewing his own and Jane Goodall's studies, he observes that older chimpanzees can actively inhibit most signs of emotions if they so choose. This means that a chimp who has discovered a source of food in the presence of a stranger or a dominant chimp does not rush over for a quick meal. Instead, 'he waits as long as he has to until the coast is clear; or he might even get up, lead the other animal somewhere else, and then (while the follower is otherwise engaged) circle back for the hidden food'. Menzel is certain that the other chimps in such a group sometimes assess this deceptive behaviour with uncanny accuracy: 'The less eager and obvious he tries to act, the more closely the followers might keep him under surveillance'.

The studies of Frans de Waal (1989) on various primate groups in zoos make entertaining and enlightening reading. It becomes very difficult not to conclude that higher primates have developed levels of social intelligence which approach our own.

In the laboratory, as we become more proficient in communicating with chimpanzees, it becomes apparent that they are capable of a much wider range of intellectual skills than was previously thought. We will deal briefly with the ability of bonobos to engage in abstract linguistic processing in a later chapter on the origins of human speech. It would be worth, however, including here a few observations of some of the investigators who have worked closely with chimpanzees. Kortlandt (1973) became deeply impressed by watching 'Washoe', a chimpanzee trained by Gardner and Gardner (1969) to sign (using ASL – American Sign Language – devised for the hearing-impaired), perusing an illustrated magazine. As with a young child, when Washoe saw an advertisement for vermouth, she signalled 'drink'; when spying a picture of a tiger, she signed 'cat'. As Kortlandt observed: 'It was fascinating to see a chimpanzee "thinking aloud" in gestural language, but in perfect silence, and without being rewarded for her performance in such a situation'.

Rumbaugh and Savage-Rumbaugh (1978) also point to a phenomenon which appears to occur some time during the course of training: the feeling that one has started to communicate, even if in a limited way, with the chimpanzee. Part of this feeling follows unexpected word usage and linguistic innovation on the part of the ape – the sense that the chimp is experimenting with language to explain something to the trainer.

In any case, our efforts to communicate with chimps under laboratory conditions are strictly a human-centred activity within a human-created artificial environment – expecting the animals to engage in quasi-human activities. We set the rules of the game to our advantage – sometimes we even move the goalposts. Whether or not chimpanzees, or other animals, operating in their own natural habitats, possess a 'real' language as linguists would define it is not as important as the recognition that animals may possess extremely sophisticated systems of communication, systems which involve a great deal of highly complex information processing on the part of both sender and receiver. As Menzel concludes: 'Chimpanzees perceive the world and interpret each other's behaviour in ways that are not ridiculously different from the ways that we ourselves use, especially when we are silent and non-sedentary'.

As we develop increasingly sophisticated methods, both our unjustified anthropocentrism and our justified scientific scepticism will recede as we rediscover what Charles Darwin already understood over a century ago: the differences in intelligence between us and higher animals are differences of degree, not of kind.

Sub-Human Self-Consciousness

The foregoing provides evidence that the higher apes engage in advanced intelligent behaviour. Almost everyone agrees that these animals are intelligent. Few, however, are prepared to grant that their level of intelligence is sufficient to allow one to ascribe to them the property of self-consciousness. However, this once again demonstrates our anthropocentric prejudices. As the primatologist Russell H. Tuttle, Professor of Anthropology at the University of Chicago, has emphasized: 'Of all the arrogant conceits in which humans have indulged, among the greatest must be counted the claim that only we are aware of ourselves, have individual identities, and think about what we are doing, have done, and will do' (Tuttle, 1990).

In the present work, *self-consciousness* is defined as *a special form of intelligence which is able to perceive a discontinuity between the system itself and the rest of the universe. As such, the system perceives a uniqueness about itself, i.e. it recognizes itself.*

By this definition, chimpanzees clearly possess self-consciousness, whereas monkeys may or may not. This assertion is based on the behavioural response of animals seeing their own reflection in a mirror. Certain species of monkey, when catching a glimpse of themselves in a mirror, respond as if they see another strange monkey. They may engage in threat behaviour. Some appear never to learn: no matter how frequent the encounter, the self-image is treated as another, intruding monkey. Chimps, on the other hand, both in the wild and under controlled experimental conditions in the lab, are puzzled by a mirror. In due course, they learn to recognize themselves. The most convincing experiment demonstrating this recognition is as follows: a mirror-wise chimpanzee is anaesthetized. While unconscious, a red patch is painted over an eyebrow or an ear tip, where the ape cannot see it without the aid of a mirror. After recovery, the chimp is returned to its quarters with access to a mirror. Upon seeing its face reflected in the mirror, it starts to explore the red patch: rubbing the spot and then smelling its fingers. It returns to the mirror repeatedly to examine the changed image (Gallup, 1975, reviewed by Tuttle, 1990).

A fish, such as a grouper, exhibiting strong territoriality, will continue to attack its self-image in a mirror until it has broken the mirror. It may turn out, as we devise other tests, not using mirrors, that monkeys possess a self-image somewhat intermediate between a fish and a higher ape.

Other students of primate sociobiology, such as Donald R. Griffin, consider that: 'When animals devote elaborate and specifically adjusted activities to communication... it seems rather likely that both sender and receiver are consciously aware of the content of these messages' (Griffin, 1984a,b). Griffin cites earlier workers, such as A. Jolly and N. K. Humphrey, who suggest that consciousness arose in primate evolution when societies developed to the stage where it became crucially important for each member of the group to understand the feelings, intentions and thoughts of others. To become 'natural psychologists', they needed to acquire internal models of the behaviour of their companions. It would seem inevitable that, sooner or later, comparison between others' and self states would lead to a clear sense of personal identity. Griffin suggests that such a process may occur in non-primate species as well; that is, that it could apply to any animal living in mutually interdependent social groups.

Only a strident anthropocentrism would keep us from recognizing that humans are not unique in achieving states of consciousness which lead to self-recognition. If it turns out (when all the evidence is in) that we do it better, or more profoundly, then that would merely reinforce the thesis that intelligence, in its myriad manifestations, comprises a broad *spectrum* of phenomena.

Instinctive Behaviour

Most people do not consider 'instinctive' behaviour as truly 'intelligent'. This is not an unreasonable position – clearly, the two behaviours, instinctive and learned, are not on a par. However, both are part of a spectrum of intelligence. Both fulfil the criteria for intelligence posited earlier. That is, instinctive behaviour, although it may not involve higher forms of thinking, involves information processing, and under normal ecological circumstances has as its outcome an increase in the probability of both survival and reproduction.

A fly, unlike a bee, seems to possess little or no learning capacity. Its behaviour appears to involve merely a series of reflex responses (Reichardt and Poggio, 1981). Yet when it avoids being swatted, or lays its eggs in a mouldering carcass, it is engaging in intelligent behaviour – even if it is unable (presumably) to 'understand' the significance of its actions. Its intelligence is so limited that it appears unable to alter its responses when the situation calls for it. For example, in nature a fly is unlikely to be attacked from two different directions simultaneously. Swatting a fly with two hands coming from two different directions simultaneously so overloads a fly's nervous system that it almost invariably gets killed.

A spider has programmed into its brain the information for weaving a beautiful, intricate web, characteristic for that species. This information is stored in its brain and is inherited. One is tempted to suggest that rigid behaviour patterns which are inherited are analogous to the read-only memory (ROM) devices of computers, while those involving learned behaviour are analogous to random-access memory (RAM) devices. An intermediate behaviour pattern, such as the imprinting phenomena observed among higher vertebrates, would correspond to programmable read-only memories (PROMs). Such a concept would imply that the evolution of the brain would have involved an increase in the number of ROMs and PROMs, followed by an even faster increase, in higher mammals, of RAMs. However, this is only a metaphor: the architectures of present-day computers and the human brain are totally different. We will compare computers and human brains in later chapters.

The realization that instinctive behaviour involves a genetic hard-wiring of the brain (to continue using the computer metaphor) allows us to differentiate between two kinds of learning. The first is what we usually think of as learning: for example, a rat learning to find its way through a maze to find its proverbial piece of cheese. The second is more subtle. It involves neither the nervous system nor the individual. It is trans-generational and it involves storing the learned information in a species' DNA. For the spider, spinning inefficient webs which catch nothing means that the spider is not likely to reproduce. The web-spinning program transcribed from the information stored in its DNA is a failure and that particular DNA will not pass its information on to future generations. Successful web-spinning programs, on the other hand, will be rewarded by the successful reproduction of the DNA transmitting such a program. There is no difference, in principle, between a rat eliminating false moves and storing the successful one in its brain, and a species of spider eliminating false webs and storing the successful ones in its DNA. If learning is defined, as we did earlier, as the ability to modify intelligent activity on the basis of previous experience, then both the rat, as an individual, and the spider, as a species, engage in a true learning process. We shall amplify this concept shortly.

Are Brains Necessary for Intelligent Behaviour?

Lower invertebrates lack brains. They do, however, possess nervous systems capable of analysing their environment and responding to it. Furthermore, invertebrate nervous systems may exhibit learning behaviour, although it involves only that most primitive of all learning behaviours – conditioned reflexes.

The California sea snail *Aplysia* has a nervous system consisting of a mere 18 000 neurons (the human brain has about a million times as many). Yet *Aplysia* can learn (Blakemore, 1988; Marcus and Carew, 1990). The slug has a gill which it pushes out into the water. If the gill is touched, it activates a reflex mechanism consisting of 24 sensory nerves plus six motor neurons which instruct the muscles to contract, thereby causing the gill to be rapidly retracted. If the gill is touched lightly, but repeatedly, the slug gradually stops retracting it – the animal has been 'habituated'. However, if the slug is poked elsewhere, or the gill is insulted with a strong chemical, the reflex reaction immediately returns to its former strength. Like Pavlov's dogs, trained to salivate at the sound of a bell, *Aplysia*'s gill can be taught to retract in response to some other non-specific stimulus. That is, *Aplysia*'s reflex can be trained to become a *conditioned* reflex. One obviously does not need a brain to engage in simple forms of learning.

Is a Nervous System Necessary for Intelligent Behaviour?

When it comes to ascribing intelligence to other organisms, many information scientists will draw a line at this point: they may concede that simple invertebrates possess intelligence by virtue of the facts that such creatures possess a nervous system, that they are motile and that they can *learn*. As we move to single-celled animals, things become fuzzier, but in any case, plants are definitely out. So thoughtful a roboticist as Hans Moravec states that: 'We need make only the grossest comparisons of the plant and animal kingdoms to appreciate the fact that mobile organisms tend to evolve the mental characteristics we associate with intelligence, while immobile ones do not' (Moravec, 1988, p. 16).

Some years ago there appeared a book entitled *The Secret Life of Plants* (Tompkins and Bird, 1973). This book was a maddening mixture of scientific fact and fiction. It was unfortunate that so much folklore and misleading material was included in the book because it has been known for a long time that plants exhibit a wide range of sophisticated behavioural adaptations. For most behavioural scientists, however, the life of plants is indeed a great secret.

The term 'behaviour' is usually defined as 'response to stimulus'. 'Instinctive behaviour' is usually considered to involve a 'biologically programmed response to stimulus'. By such a definition plants engage in 'instinctive behaviour'. Even though plants lack a nervous system, they have a variety of senses by which they analyse their environment and then, on the basis of that analysis, respond intelligently using a variety of effector mechanisms which have been genetically programmed.

Furthermore, what is frequently overlooked is that some plants are able to move very rapidly in response to certain environmental stimuli. The rapid transmission of information from the site of contact to the site of action approaches that of animal responses. The Venus Flytrap traps flies and other insects to supplement its meagre nitrogen supply in the sandy soils of North Carolina. It does this by rapidly closing its opened leaf in response to objects moving on its upper surface.

The detection system of a Venus Flytrap allows it to discriminate between a piece of dirt falling on its leaf and an insect moving about. It takes energy to open up the leaf and set the trap. Evolution has selected for a system which does not waste energy on a piece of dirt. The detection and catching of insects makes this organism appear to behave more like a predatory animal than a plant. It is a plant nevertheless, and manages to catch insects without recourse to a nervous system.

Similarly, the Touch-me-not collapses its leaves and stems in response to injury or mechanical stimulation, such as cattle browsing. The speed and distance that the impulse moves along the leaves, twigs and stems is determined by the intensity of the stimulus. Again, this plant also has nothing which resembles the neurons of an animal's nervous system. The transmission of information involves a different mechanism. Although not as dramatic, other plants also exhibit movements: among the best known of these is when the heads of sunflowers turn to face the Sun – facing east in the morning and west in the evening.

The problems of survival for plants are quite different from those of animals. Plants manufacture their own food. With the exception of the two examples cited above, the need to move information rapidly from one part of the body to another – a prerequisite for achieving locomotion – is not important to plants. Much of the exchange of information among the various parts of the plant's body is achieved by hormones – the chemical messengers. Unlike transmission mediated by a nervous system, which is able to move information from one end of the animal body to another in a fraction of a second, plant hormones involve a time-scale of somewhere between minutes and weeks. Furthermore, plant hormones do not trigger specialized tissue, such as muscle tissue; instead, movement in plants involves the enlargement or multiplication of cells, a process which involves a time-scale ranging from hours to years. It is the lack of tissue specialized for rapid response such as nerves and muscles – tissues which allow animals to engage in behavioural responses as fast as the blink of an eye – which makes plants so slow and ponderous in their behavioural responses. This slowness, however, should not blind us into believing that plants are not able to engage in instinctive – that is, biologically programmed – behaviour. A few examples will illustrate the point.

Plants are *geotropic*. That is, just as some animals have been programmed to go under the ground (e.g. moles) and others up into the trees (e.g. gibbons), so the vast majority of plants have been programmed to send their roots down and their shoots up. The gravity detectors consists of tiny grains – cell organelles called statoliths – which are found in cells at the root tips and shoot tips. These statoliths tend to sink to the bottom of cells and bring about a redistribution of growth hormones by affecting the synthesis, transport or destruction of hormones in the local region. These hormones affect primarily cell elongation just behind the tip. In roots, elongation of the upper surface causes the tip to bend downward. In shoots, the opposite occurs as the distribution of hormones causes the lower surface to extend – forcing the growing tip upward.

Plants are *phototropic*: shoots are positively so, and therefore grow toward the light; roots exhibit negative phototropism – growing away from light. As with geotropism, phototropism relies on cellular organelles near the tip which act as photoreceptors and which again affect the distribution of growth hormones by altering their local transport, synthesis or rate of destruction. Incidentally, at the molecular level, the transport of organic substances, such as hormones, and of inorganic ions, as well as many other aspects of basic cellular physiology, are very similar in plants and animals – sometimes virtually identical.

Some plants have evolved special organs to facilitate the response to light and gravity. The *coleoptile* in grasses is such an organ. Grasses are among the most advanced of land plants from an evolutionary point of view. The seeds of grasses, such as oats, are relatively small, with a limited supply of stored food. Anything which helps the germinating seed to push its shoot up rapidly through the soil towards light would aid survival. Although a shoot needs to grow, its main function is photosynthesis – a function denied the plant underground. By encasing the young and vulnerable germinating shoot in a protective sheath, the coleoptile, which is able to detect very low intensities of daylight and is equally sensitive to gravity, the germinating seedling is assured of an optimum behavioural strategy for raising its shoot. Once above ground, the leaf breaks through the coleoptile to achieve its photosynthetic function.

The oat coleoptile is richly endowed with both gravity and light detectors, and produces large quantities of growth hormones. It was, in fact, while working with the oat coleoptile that Fritz Went in the 1930s first provided unequivocal evidence for the existence of plant hormones – a classic piece of work which can trace its lineage of experimentation and observation back half a century to Charles Darwin.

The grasses illustrate another behavioural strategy, adapting them to an environment which includes grass-eating herbivores: instead of growing from the top down, as most plants do, they grow from the bottom up. That is, their growing centres (meristems) are at the bottom of the shoot (rather than at the tip), thereby minimizing damage from grazers (or, for that matter, from lawnmowers).

Another case of biologically programmed behaviour in plants is the ability of many plants to flower at the same time of year, thereby ensuring reproductive efficiency. This example parallels the instinctive behaviour of many animals which engage in unlearned – that is, biologically programmed – mating behaviour during certain specific seasons of the year. In many instances, the mechanism appears to be virtually identical in plants and animals: both plants and animals have mechanisms for detecting the appropriate season by measuring the length of day/night cycles. An induction period of a sufficient number of days of the right length alters the hormonal status of the organism. This basic mechanism, then, results in a wide repertoire of behaviour: male stickleback fish develop red bellies, build nests and become extremely aggressive; red-ruffed grouse males stake out territories of heather which they defend vigorously; female deer go into oestrus, while their potential mates, having grown large antlers earlier, have been bashing each other about for some time. Plants merely flower.

It is easy for students of behaviour to overlook plants. Yet the photoperiodic detection mechanism in plants is exquisitely sensitive: in a greenhouse of tobacco plants, a night watchman checking his station, turning on a forty-watt bulb in the middle of the night for only a few minutes, will prevent those plants from flowering.

Finally, one must emphasize that whereas most behaviour in *plants* is dependent on hormones, hormones also play a crucial role in the behaviour of animals as well. Returning to the molecular dimension, it is probably no mere coincidence that one of the most important of plant hormones, indole acetic acid, is a close chemical relative of the neurotransmitter serotonin. Probably both substances are concerned with the transport of crucial ions across membranes.

Both the biological microworld of comparative molecular chemistry and the macroworld of ecology provide ample evidence for the need to extend traditional concepts of intelligence and behaviour to embrace *both* kingdoms. On the other hand, what *individual* plants cannot do is *learn*. With the exception of those discussed above, plants lacking a nervous system exhibit no rapid reflexes. Lacking reflexes, they are not capable of that most primitive of all learning phenomena – conditioned reflexes. It is probably for this reason that most scientists currently consider plants as lacking all intelligence.

In the present work, we will consider self-organizing systems which *learn* as representing more advanced information systems than those that do not. Keep in mind that many people do not consider sea slugs as possessing intelligence. For them, only animals which exhibit consciousness possess intelligence. In our scheme, animals which can recognize themselves represent higher levels of intelligence than those which cannot. In line with this thinking, organisms which cannot learn are not relegated to the 'non-intelligent' category, but instead are classed as possessing a more primitive level of it. Furthermore, as discussed above, the *collective* intelligence of any given species – plant, animal, or microbial – *is* capable of learning in the traditional sense. New information about the changing environment is acquired by the species as a whole: selecting from among a series of hit-and-miss mutations, the forces of evolution select those which are adaptive to the new environment and then store this new information in the genetic material of each species. This matter will be discussed in greater detail shortly. However, it is a powerful argument for considering that *all* bio-systems possess some measure of intelligence.

The Collective Intelligence of Microorganisms

Traditionally, bacteriologists have been inclined to view bacteria as pure colonies of single cells growing on agar in Petri plates. In the real world, however, bacteria live in complex multicellular communities (Shapiro, 1988). Some communities may consist of a mixture of many different species. Others may consist of a single species forming complex colonies. Myxobacteria, for example, never exist as single cells. Even when adverse conditions force them into dormancy, the colonies form multicellular cysts which germinate to yield a ready-made colony of thousands of individuals.

As the colony grows and the population becomes more dense the individual bacteria migrate. However, they do so in a coordinated fashion, secreting trails of extracellular slime which mark the highways along which thousands of cells migrate, in colony-coordinated rhythmic pulses. Part of the colony differentiates to form cysts; in some species colonies form elaborate, psychedelic fruiting bodies. When the colony migrates, it does so as an intact unit – any cell which moves a few microns beyond the edge rapidly returns, to be enfolded once again by the colony.

These various activities are under genetic control. *Myxococcus xanthus* has two different genetic systems for regulating movement: an A (adventurous) system which allows individual cells to move, and an S (social) system which regulates the movement of groups of cells. A defect (a mutation) in either system leads to an abnormal colony structure. If both systems become defective, the cells cannot spread at all. At least 23 different genetic factors have been identified in the A system and at least

10 in the S system. When two motility-defective mutants are combined in the same Petri dish they regain motility as long as the two mutant types remain closely associated.

Similarly, the organization of colonies into distinct concentric zones by other species of bacteria involves the formation of rings, not at specific positions on the substrate on which they are growing, but at specific times during the growth of the colony. Thus, both biological clocks and the timing of development are important features of colony growth, just as they are in the growth of multicellular organisms.

Bacteria as a group have been around for a few thousand million years. They are so widely dispersed on this planet that some scientists refer to them as the 'global organism' (Sonea, 1988). One reason for the universal success of this apparently vulnerable, microscopic form of life is that bacteria have evolved a number of mechanisms for exchanging genetic information. This occurs not only among members of the same species, but also across different species, using viruses and plasmids as genetic tools. This allows bacteria to gear up rapidly to new metabolic opportunities or to adapt to new requirements imposed by the environment.

If one expands the concept of 'behaviour' to include metabolic behaviour, then clearly the bacteria behave differently as a result of their changed genetic status. This changed genetic status is a reflection of the organism's environmental experience. Thus, if we define 'learning' in traditional psychological terms - an enduring change in knowledge or behaviour resulting from experience - then the bacteria as a totality are capable of learning.

To sum up, therefore, a single bacterium represents a biological system near the bottom end of the intelligence spectrum. However, the conglomerate - all the bacteria and their genetic transfer mechanisms together - represents a biological system with a much higher level of intelligence than any one individual bacterium. This ***collective intelligence*** of bacteria - possessing a much more sophisticated system for analysing the environment and then responding intelligently - comprises a manifestation of an information-processing system whose combined intelligence may be considered to be substantially greater than that of any one individual ***multicellular*** organism.

DNA as Memory

The idea that an entire species may have a memory (stored in its DNA) which provides a mechanism for a species, as a whole, to engage in learning behaviour is so important a concept that it deserves amplification. Let us take the example of a species of insect 'learning' to overcome a new environmental threat: the application of insecticides by farmers (Taylor *et al.*, 1995).

The tobacco budworm (*Heliothis virescens*) is a major insect pest on several crops, in particular cotton. The budworm can be chemically controlled by means of 'pyrethroid' insecticides, which exert their effect by interfering with the normal functioning of the voltage-gated sodium channels in nerve membranes. Unfortunately, in some parts of the country, pyrethroid-resistant strains are beginning to emerge. Resistance to the insecticide is genetically linked to a DNA marker for a sodium channel locus called ***hscp***. Choosing five geographically widely separated populations of the insect, Taylor and colleagues compared the ***hscp*** locus with another genetic locus which had no relation to pyrethroid insecticide resistance and which, therefore, could act as a control locus. Earlier studies by other workers on the North American budworm population, ranging from Mexico to Georgia, showed that, overall, the species showed little genetic differentiation. Therefore the Taylor study would be able to discern if a particular genetic alteration (such as a mutation) conferring insecticide resistance was being selected for. This did, in fact, turn out to be the case.

The acquisition of insecticide resistance is a classic example of a ***species*** learning how to overcome an environmental threat while the individuals making up the species are virtually oblivious to it. At the moment there is no evidence that the individual budworms avoid ingesting the poison: they are too dumb to know it is there, or that it is bad for them. On the other hand, individual variants whose voltage-gated sodium channels are modified so as to cope better with the pyrethroid toxin are selected for. Thus the species' DNA is reprogrammed, and the species' future behaviour – in this case, its metabolic behaviour – has been altered as a result of its historical experience.

In principle, there is no difference between a laboratory rat learning to engage in behaviour which avoids an electric shock and the budworm species learning to find a way of not succumbing to an insecticide. In the rat brain, the neural pathways which recorded unsuccessful behaviour patterns tend to fade while the neural pathways which represent successful avoidance behaviour become selected for. Likewise, the DNA of unsuccessful poisoned budworms is eliminated while the DNA of those individuals who could cope successfully tends to be reproduced. In both instances – the individual rat and the budworm species – past experience leads to adaptive alterations in future behaviour.

Several decades ago, Norbert Wiener (1961), in discussing these two types of learning, referred to them as ***ontogenetic*** and ***phylogenetic***. In the biological sciences, the term ***ontogeny*** refers to the developmental history of the individual while ***phylogeny*** refers to the evolutionary history of the species or other taxa (genus, family, order, class or phylum). Wiener did not have the wealth of information available to us today, but he quite correctly pointed out that higher mammals and, in particular, humans, tend to engage in ontogenetic learning, while birds have much of their behaviour governed by phylogenetic learning.

The fact that all known species – from bacteria to elephants – are able to engage in phylogenetic learning demonstrates that all living organisms exhibit the capacity for learning – if not as individuals, then as a species. Even the DNA of viruses is able to learn, a matter of considerable practical importance to medical clinicians and animal and plant pathologists. At this point in history, the bane of AIDS research, for example, is the ease with which the virus is able to mutate. In so far as the capacity to learn must be considered a manifestation of intelligence, we may conclude that all life forms must possess some measure of intelligence.

Summary and Concluding Remarks

Intelligence is a property which emerged in advanced IPSs. Initially it became a device for adapting biosystems to their environment. Ultimately, it evolved in humans (and to a lesser extent in some other animals) into the ability to control the environment for the benefit of the organism. That is, the environment, rather than the organism, became modified.

Intelligence can be ascertained only by observing the behaviour of an IPS. The system must be able to analyse its environment and respond, on the basis of that analysis, in a way which enhances its survivability and reproducibility. Another property of intelligent IPSs is that the system is able to learn.

An analysis of intelligence makes plain that we are dealing with a spectrum of phenomena. At the bottom end is 'proto-intelligence', comprising systems which are on the borderline of intelligence. Next come single-cell organisms, capable only of trans-generational (phylogenetic) learning. Plants fall into that same category, but exhibit more sophisticated methods of analysing their environment and responding to it. The sea slug *Aplysia* has a nervous system but no brain. However, it exhibits the capacity for ontogenetic learning in that it is capable of conditioned reflex reactions to its environment. A fly has a brain which reduces the chances of it being swatted, and which enables it to find food and search for places favourable for laying its eggs. A bee brain is more advanced: it shows that this organism has an abstract version of a map of the world in its head, and that it can communicate features of this world to other nest mates. The vertebrate brain has evolved to show increasing sophistication, until, at the upper, mammalian, end we find clear indications of consciousness and self-consciousness. At the very top is the 'collective intelligence' made up of the social organization of intelligent individuals.

Difficult to place in this spectrum is something which might be called 'ecosystems intelligence', exemplified by the entirety of the microbial world covering this planet.

In any case, emerging from this analysis of intelligence is the concept of *hierarchy*. Both the phenomenon of *information* and the phenomenon of *intelligence* exhibit hierarchies of complexity. That is, complexity builds on pre-existing complexity, a matter which becomes self-evident as we study the evolution of the mammalian brain. Without this understanding – that there exist many levels of complexity which can be graded into a hierarchy – it becomes impossible to develop either a general theory of information or of intelligence.

The next three chapters are meant to explore the emergence of intelligence, to help make comprehensible how such a strange phenomenon – contra-indicating all that we know about the present-day laws of physics – may have arisen, and, for that matter, is arising today.

· 5 ·

Proto-Intelligence

Introduction

...you can never get hold of the real beginning of anything.

Teilhard de Chardin (1956, p. 64).

Proto-Intelligence

If the phenomenon of intelligence is to be viewed as a spectrum of phenomena, what is the underlying mechanism which created such a phenomenon in the first place? What lies at the base of the phenomenon of intelligence? All living systems, including the subsystems of advanced biosystems, exhibit some level of intelligence as defined in the present work. The question then must be asked: may inorganic, non-living systems exhibit intelligence?

Leaving aside both machine and electronic intelligence, which will preoccupy us in a later chapter, let us consider the case of a crystal of manganese dioxide dropped into a solution of potassium permanganate. The crystal, instead of dissolving, becomes the focus of an autocatalytic reaction which converts the solution of potassium permanganate into more manganese dioxide.

The ability of a crystal of manganese dioxide to convert its external environment into more of itself demonstrates that an inorganic system may reproduce. As such the crystal has fulfilled one criterion for ascertaining intelligent behaviour – the enhancement of reproducibility.

A seed crystal of salt dropped into a supersaturated solution of salt or a seed crystal of silicon dropped into a cooling mass of molten silicon will also trigger reactions which result in the precipitation of salt crystals or the growth of the seed crystal into a large crystal of silicon. Again, the criterion of reproducibility has been fulfilled. Does this mean that crystals possess intelligence? We are now at the borderline of our spectrum.

By our definition, individual crystals do not fulfil the criterion for true intelligence. Under usual circumstances, crystals fail to duplicate themselves. When they do reproduce, they are engaging in intelligent behaviour. However, once a reaction is completed, crystals return to their wholly inanimate state, totally subjected to the vagaries of fate which their environment may impose on them. Nevertheless, if only briefly, crystals may exhibit 'flashes of intelligent behaviour'. For this reason we may invoke the concept of *proto-intelligence.*

The concept of proto-intelligence is important: it is important in its own right. It is also important for any analysis of intelligence, in general, and electronic intelligence in particular. *Proto-intelligence* may be defined as *phenomena which involve aspects of intelligent behaviour but which exist only temporarily, or if permanently, only partially satisfy the criteria for intelligence.*

For example, memory is an integral part of the learning process. Many non-living systems exhibit the property of memory: 'memory metals' and other materials 'remember' previous shapes and processes. Similarly, a disturbed pendulum returns to its resonant frequency; and, of course, computer systems exhibit memory. All these constitute a form of proto-intelligence.

Still more primitive in organization, involving fewer components, are molecules, atoms and subatomic particles. However, as Klaus Haefner (1991) has pointed out, these entities manage to maintain their identity – a proton behaves as a proton, an electron as an electron, and a hydrogen atom as a hydrogen atom. It is the ability of these entities to maintain their physical integrity, to engage in both information exchanges and information processing, which implies that they are stable information systems which exhibit aspects of intelligence. However, both their inability to learn and their inability to reproduce themselves put these inanimate entities into the category of proto-intelligent systems.

The concept of proto-intelligence must be an integral part of any discussion which concerns itself with the origin and early evolution of intelligence. As with any evolutionary process, a major shift in organization is preceded by new features in existing systems which set the stage for such an evolutionary discontinuity. Sometimes the combination of these new features creates an intermediate, such as the appearance of amphibians in the evolution of fish into land animals. In fact, this particular evolution may serve as a conceptual model for the evolution of non-intelligent information systems into intelligent ones.

The evolution of lungs in animals living on dry land was preceded by, and derived from, the evolution of swim bladders in fish. The evolution of bones, important for survival on land, was preceded by the evolution of bony fish. Fish don't need hard bones to function well in water. Sharks, which only possess relatively soft cartilage, have fared very well for hundreds of millions of years swimming around in the oceans. Fish moving upstream into fresh water, however, faced an uncertain supply of vital calcium. What more logical solution than to create a calcium 'bank' by depositing calcium compounds among the cartilage? Thus the cartilaginous mechanical structures may be considered as 'proto-bones'; similarly, a swim bladder can be considered to be a 'proto-lung' and certain fins of fishes as 'proto-legs'.

The gap between animals which spend their life on land and those which spend their life in water is bridged by the amphibians. Amphibians, such as frogs and salamanders, represent an intermediate state. In like manner, proto-intelligence represents an intermediate stage in the evolution of simpler information systems into complex, intelligent systems. We look for phenomena which, in themselves, do not satisfy our criteria for intelligence, but which are related. We would not look to the eye of a fish to evolve into a lung, nor its blood vessels to become bones; instead we look to its swim bladder and its cartilage. Similarly, we look to information systems exhibiting limited aspects of intelligence as constituting the phenomena of proto-intelligence from which more advanced intelligent systems may have evolved.

The above discussion has identified several phenomena which may be classed as proto-intelligent: (1) survivability, as demonstrated by the stability of atomic particles; (2) reproducibility, as exemplified by the growth of a crystal; and (3) memory. Other phenomena will be identified as we delve more deeply into the subject. However, one form of proto-intelligence, which is so basic as to make impossible any informed discussion of intelligence without it, is *feedback*. All intelligent systems process information along at least one major feedback loop.

Feedback loops have their antecedents in the form of regular cycles which, under certain circumstances, may represent a form of proto-intelligence. Such cycles may be observed in the organization of an atom, a swinging pendulum, a resonating electronic system, a planet swinging around a star, the stability of Bénard cells or cyclic chemical reactions – that is, any cyclic phenomenon with a regular periodicity involving two counteractive forces: centrifugal/centripetal, electrostatic/electromagnetic, oxidative/reductive etc. Like a gyroscope spinning, regular oscillations tend to maintain the stability of the system even when the environment changes. It represents a major mechanism in the survivability of a system. In some proto-intelligent systems, the oscillations dampen down in time and finally disappear. The system collapses. In others, such as atoms, the organization is maintained indefinitely by a complex interaction of internal forces. In contrast, in bio-systems, internal rhythms are maintained by the controlled inputs of an external source of energy. Plants do it by utilizing sunshine, animals by eating plants or other animals. Similarly, mechanical or electronic systems function by having available a source of energy.

All intelligent behaviour involving the interaction between a system and its environment is based, at least in part, on an exchange of information involving feedback loops.

A grandfather clock is a prime example of a piece of machinery which constitutes a proto-intelligent system. It is a goal-oriented object. Its goal (imposed by its human designer) is to move the hands of the clock in small, even, exact and continuous steps. It processes information in that the time it takes for a weight to descend is converted into the movement of the hands of the clock. It achieves this goal by regulating the input of energy by means of mechanical gears and levers and a steadily swinging pendulum, so as to achieve a constant output. However, the clock contains no system which compensates for changes in the environment. The steady, gravitational pull of the weight moves the steady, ticking machinery. When the weight reaches the end of the chain, the clock stops. If the weight becomes too heavy, the clock tries to run faster and may break. An increase in temperature causes the pendulum to become longer, and the clock slows down. The reverse happens if the outside temperature cools.

The grandfather clock exhibits aspects of intelligence: it is goal-oriented. It processes information. It regulates (in limited fashion) the throughput of energy. It converts energy into information. And its pendulum exhibits memory (in so far as it will return to its basic frequency of oscillation if disturbed). However, it is totally dependent on the right combination of externally imposed factors in order to achieve its goal. And it cannot learn. For this reason, although it has many of the attributes of intelligence, a grandfather clock should be classed as a proto-intelligent system.

We will discuss in a later chapter how another proto-intelligent system, the home thermostat, may evolve into a form of primitive mechanical intelligence, capable of learning. It should suffice here to point out that the evolution of intelligent systems from proto-intelligent systems must have occurred on numerous occasions, involving quite different kinds of advanced information-processing systems. The polyphyletic origin of intelligence further confounds the taxonomic problem of defining unequivocally what is a proto-intelligent system and what is an intelligent one. At the bottom end, the intelligence spectrum is blurred. It becomes impossible to create a clear demarcation between information-processing systems which exhibit proto-intelligence and those which could be classed as truly intelligent.

Some Further Thoughts About the Preconditions for Intelligence

The most basic assumption one can make about the phenomenon of intelligence is: **intelligence is a property of advanced information-processing systems.** That is, just as information is a property of the universe, so is intelligence a property of evolved information-processing systems.

A system contains information if it exhibits organization. The more information a system contains, the greater and more complex will be the organization of that system. Advanced information systems exhibit many complex organizational features, some of which set the stage for proto-intelligence. Of these, **five basic processes** can be identified as being prerequisites for establishing complex proto-intelligent and, ultimately, intelligent systems. These are: (1) **resonance**, (2) **feedback**, (3) **duplication**, (4) **differentiation** and (5) **recombination.**

1. *Resonance.* A major feature of many organized systems is that they exhibit resonances. Resonances produce oscillations. These result in regular cycles, ranging from electron spin resonances in atoms to biorhythms in humans. Regular cycles favour *stability*: like a spinning gyroscope, it becomes much more difficult to displace or distort a cycling system than a static one. However, a resonating system also allows minor *variations*. Each recursive cycle permits the introduction of variations into the cycle. The more complex the system – the more numerous and complex are the cycles and sub-cycles – the easier it becomes to introduce additional variations.

If 'variation' represents the first half of the mechanism of evolution, 'selection' represents the second half: systems which work, survive; those which do not, do not. Resonating systems set the stage for feedback loops. Such feedback loops arise not only between the various components of a system – thereby assuring their organizational stability – but also between the system and its *environment*. The system begins the rudimentary task of detecting changes in its environment and responding to it. This is the birth of intelligence.

2. *Feedback.* Feedback loops are a prerequisite of intelligence. Feedback loops stabilize *complexity* and allow a system to interact with the environment in a way which enhances *survivability.*

A simple example will suffice: the shoot of a tomato plant produces certain hormones and vitamins which are transported to the root tips. Without these substances the roots cannot grow. Conversely, the root produces other hormones and amino acids, vital for healthy shoot growth. Damage to one part of the plant, either root or shoot, causes the other part to 'wait' until the damage is repaired and normal balanced growth and development can, once again, resume.

The growth of a tomato plant exemplifies the role that feedback loops play in maintaining the organization and integrity of a complex system. Such organization and integrity are wholly dependent on the ***interdependence*** achieved between the various parts making up the system.

3. *Duplication.* One of the safest ways of introducing variations into a system without destroying it is for the system to duplicate – to divide (or multiply) it into many identical units, and then, and only then, 'experiment' on some of these sub-units. After such cloning, a variation lethal to one of the multiple units merely constitutes a failed experiment. It need no longer be lethal to the whole system.

4. *Differentiation.* Duplication, followed by a process of variation affecting the individual (duplicated) parts, leads to differentiation.

This is how insects evolved: from a multi-segmented arthropod such as a millipede in which all the segments have legs and are generally similar, to an insect in which only the middle section, the thorax, has legs. Other segments became altered so that, instead of legs, insects produced antennae or wings. As the vertebrae in our backbone attest, we humans can also trace our ancestry back to segmented animals.

If each segment, initially similar, evolved to carry out a specialized function *within* an individual organism to allow a higher survival value, then a beehive illustrates that the same may happen *across* individuals. The principle of differentiation (as with the other principles) may occur at any level of organization. A beehive provides an example of differentiation in social systems. More primitive insects consist of individuals which, except for the sexes, are identical. Bees, on the other hand, evolved societies in which classes of individuals became *anatomically* differentiated in order to carry out specific functions. In addition to being physically different, there are further behavioural differences: for example, workers involved with the care of larvae carry out tasks different from those workers whose main job it is to gather pollen and nectar. In ant societies this differentiation of individuals evolved even further to result in several anatomically distinct castes, each adapted to carry out a specific task. In addition to the queen and the males, whose main function is reproduction, there are the soldiers who guard the nest, the minor workers who take care of feeding larvae and housekeeping, and a series of media workers who engage in foraging for food, as well as other chores (Hölldobler and Wilson, 1990).

5. *Recombination.* It is probably true to say that the process of *combination* and *recombination* is the primary source of *major* evolutionary advances (rather than the steady build-up of point mutations). One may take, as an example, the 'eukaryotic' cell, which possesses a nucleus (an organelle lacking in the more primitive 'prokaryotic' cells). It is now believed that the eukaryotic cell is derived from a combination of more simple (prokaryotic) cell types. For example, in photosynthesizing plant cells, the chloroplasts are thought to represent evolutionary derivatives of formerly independent, photosynthesizing cyanobacteria. In both plant and animal cells, the mitochondria (which comprise a cell's 'powerhouse'), similarly, represent derivatives of bacteria which in ancient times were independent (Margulis, 1970).

In like manner, the basic processes which underlie the immune response of our own bodies to invaders of the bloodstream are also the product of recombination. The millions of different antibodies which comprise part of our immune mechanism are created by combining and recombining the code of a few hundred (or, at a maximum, a few thousand) genes.

The principle of divergence and recombination is evident also during the development of an embryo into an adult individual. The human body develops from a single cell into scores of cell types – different kinds of blood cells, liver cells, muscle cells, heart cells, brain cells etc. Not only are the cells making up different tissues, organs or systems different from each other, but also *within* each tissue, organ or system different cell types exist.

In the very early stages of human embryo development, after a fertilized egg has divided only a few times, each cell appears to be a mere *duplicate* of every other cell: any isolated cell can give rise to an entire organism (the cells are totipotent). Very soon, however, as cell division proceeds, cells begin to *differentiate*, and seem to have their totipotency restricted, first to being able to produce large categories of cell types (the cells are still pluripotent), and then, as the embryo becomes more and more complex, the potency is further restricted until, finally, most cells can only produce their own cell type (unipotent), or, in some instances, cannot even divide at all, as in the case of mature red blood cells in the human circulatory system.

The process of differentiation involves the activation and inactivation of various genes. That is, the information stored in the DNA of the cell is released and processed in relatively small packets, in a tightly controlled time sequence. Which packets are activated, which are shut down, and which are inhibited from being activated in the first place (repressed), is dependent on sequences of chemical messages, some originating from the cell's own metabolic machinery, some originating from neighbouring cells, and others arising from cells at some distance from the receptor cell. This last situation involves hormones – the chemical messengers – which continue to play such a vital role in integrating the diverse cells of the adult human body into a coherent whole.

In the fruit fly, three genes encode chemical signals that specify the structures appearing along its long axis (Nüsslein-Volhard, 1996). One signal controls the development of the front half of the egg, giving rise to the head and thorax; a second controls the region which develops into the abdomen; and a third controls the development of structures at the two extreme ends of the developing larva. Therefore, each cell in the developing embryo is exposed to a gradient of chemical messages, emanating from itself, its various neighbours and more distant sources of metabolic activity. This explains the principle first stated by the German botanist Herman Vöchting in 1878: 'The function of a cell is determined by its morphological position'. That is, *where* a cell is located in the body of a plant or animal determines what kind of a cell it is going to become.

The chemical messages to which developing cells are exposed are of different kinds: some may instruct a cell to divide, others to enlarge and still others to differentiate (in some cases leading to the point of death), while a fourth kind of message may instruct a cell simply to be quiescent and not do anything. Most important are the messages which tell cells to produce or release hormones, neurotransmitters or other chemicals which themselves act as messages for other cells. These mechanisms create *feedback loops* and allow for *integration*. By being exposed to various gradients of diverse growth substances, each cell is under the control of a unique *combination* of messages. Relatively few kinds of substances, in a multiplicity of combinations, are able to create the diversity of differentiated cells which make up the mature organism.

To sum up, advanced information-processing systems needed to evolve a number of processes such as resonance, feedback, duplication, differentiation and recombination to satisfy the preconditions for intelligence. Out of these processes there could emerge systems which showed phenomena associated with intelligence: stability and self-preservation, non-destructive interaction with the environment, reproduction, memory and learning. This is, of course, what happened during the origin of life – a matter to be discussed in the next chapter.

· 6 ·

The Origins of Bio-Intelligence

Introduction

The origin and evolution of living organisms on this planet – that is, the phenomenon of ***life*** – represents the best example of the origin and early evolution of the phenomenon of true intelligence. Until recently it represented the only example. This has changed with the advent of machine and electronic intelligence – a matter to be discussed in the next chapter. In this chapter we will speculate about how pre-biotic systems might have evolved into true living forms, that is, how bio-intelligence originated.

Teilhard de Chardin's Concept of the 'Corpuscle'

The principles described in the previous sections form a constellation of phenomena and processes which constitute the prerequisites for the emergence of intelligent systems. At this point it may be helpful to introduce the concept of the *corpuscle*, as Teilhard de Chardin (1956, pp. 19–25) proposed it: complexity does not mean simple aggregation, nor the undefined repetition of units. Complexity implies 'combination' which knits together a certain fixed number of elements within a closed whole of determined radius, such as an atom, molecule, cell, plant or animal. Such a unit Teilhard calls a corpuscle. This is in contrast to units which arise by aggregation and crystallization, in which the arrangement remains incomplete: one can always add on from the outside. This form of non-corpuscular organization applies equally to a microcrystal and a giant star. There is no inherent unit, no self-limiting boundary – only an accidental 'rounding off' of the system.

This is the opposite of 'combination' which produces groups that are structurally complete. This does not preclude such groups from duplicating, but they are indefinitely extensible from ***within*** – what classical biologists call growth by 'intussusception', rather than by accretion.

Teilhard considers that the *corpuscle* is: 'a unit truly and doubly "natural" in the sense that while organically limited in its contours so far as its own existence is concerned it also, at certain higher levels of internal complexity, manifests strictly *autonomous phenomena*' [italics in original]. Although it is not entirely clear from Teilhard's treatise exactly what he means by 'autonomous', presumably it refers to internal processes which have been sufficiently insulated from the external environment to allow their operation to proceed independently of that environment. That is, the boundary provided by the corpuscle excludes the environment from directly controlling the processes internal to the corpuscle. Presumably this is as true for electrons orbiting around the nucleus of an atom (but within its atomic confines) as it is for the enzyme-controlled metabolic reactions within cells, or the heat regulating system of our own bodies which buffer such enzyme reactions.

Using Teilhard's concept of the corpuscle as our point of departure, it must be clear that to reach the levels of integrated complexity required to achieve the corpuscular state requires feedback loops. The internal organization of such a corpuscle must also have previously undergone evolutionary processes which involved duplication, differentiation and recombination – processes which at the molecular level are a part of all living organisms. This chapter will concern itself with the transition from proto-intelligence to true intelligence, i.e. with the origin of life.

Forming a Bio-Corpuscle

For a fully formed bio-corpuscle to exist there must arise some form of delimiting membrane: some structure which separates the individual, corpuscular information-processing system from the rest of the environment. Harold Morowitz, Professor of Biology and Natural Philosophy at George Mason University, considers that the formation of such a membrane system must indeed have been the first step in the origin of life on this planet. Morowitz (1993) contradicts the conventional wisdom by considering that 'cells' originated first, followed by proteins, while the genetic apparatus appeared last. The following is a summation of Morowitz's thesis.

The original 'cell' was nothing but a vesicle – a sac surrounded by a bimolecular–molecular layer which acted as a membrane separating the interior from the exterior. Morowitz postulates the presence of hydrocarbons, including long-chain alkanes and aromatic rings, floating to the surface of the oceans. These hydrocarbons, exposed to ultra-violet light, lightning and lava flow, would interact with some of the dissolved minerals in the water. The combination of the influences would lead to polar alkanes which, in turn, would lead to bimolecular leaflets with a polar exterior and a non-polar interior. Under the influence of wave action, such structures would end up as closed vesicles.

To quote Morowitz (1993, p. 173): 'This formation of closed vesicles was a major event in biogenesis: It represented the origin of supramolecular entities, three-phase systems consisting of a polar interior, a nonpolar membrane core, and a polar exterior – the environment'.

Membrane systems would allow entirely new and unique molecular environments to come into existence. Should there be created chromophores in this internal environment, then there would be the possibility for the vesicle to become energetic. Sunlight shining on such chromophores would be transduced into electrical potential, thereby establishing a foundation for bioenergetics.

Such a chromophore would ultimately evolve into a system that we see today in the bacterium *Halobacterium salinarium* (Lanyi, 1995). This bacterium has, embedded in its cell membrane, bacteriorhodopsin, a chemical similar to the visual pigment found in the human eye. Upon shining sunlight on it, bacteriorhodopsin provides the energy for pumping protons out of the cell. This creates a so-called protonmotive force by developing a potential across the membrane (Miyasaka, 1995).

A second membrane transport system would be required, this one for transporting electrons. These two systems, the electron transport and the proton transport, would set the stage for oxidation/reduction reactions and acid/base reactions, respectively. Finally, the introduction of phosphorus-containing compounds would complete the stage for modern bioenergetics.

Today, the energy flow in bio-systems consists of light energy being transduced to create a reduced compound (AH_2) and the formation of pyrophosphate and polyphosphate to create the chemical energy to drive all the rest of metabolism. This process, to quote Morowitz: '...is one of the most powerful generalizations of modern biochemistry, and there is every reason to believe that it is primitive'.

This is in line with Morowitz's insight that today's 'intermediate metabolism recapitulates prebiotic chemical evolution' (Morowitz, 1993, p. 177). However, unlike the metabolism of today's living organisms, the pre-biotic 'metabolism' involved chemical reactions which were not dependent on enzymatic activity, or even catalysis. Rather, they involved reactions natural to small molecules made up of carbon, hydrogen, oxygen, phosphorus and (possibly) sulphur.

Among other products of this pre-biotic metabolism would be the keto acids, which, in the presence of ammonia, would be converted to amino acids. These would polymerize to form proteins, some of which could achieve the enzymatic role of modern proteins. Others would become structural proteins to give further form and shape to the vesicles, while a third category would become part of the transmembrane transport system. In the long run, other specialized proteins dealing with a cell's motility, genetic apparatus, signal transduction etc. would arise to complete a modern cell's protein complement.

The introduction of nitrogen into the early metabolism would also set the stage for that other class of chemicals, the purines and pyrimidines, the basic building blocks of nucleic acids. During this time the protobiotic systems would be engaged in a process of Darwinian selection which would refine the self-reproducing capacity of the system. A fairly late addition to the evolving system would be a genetic code. Once it was in place, chemical evolution would give way to biological evolution and the 'universal ancestor' of all living things would have come into existence.

Is the Vesicular Theory Enough?

Morowitz's thesis that vesicles appeared first, proteins later and the genetic apparatus last, is appealing and may well be correct. However, there arises the question as to how, if the internal vesicle environment favoured the formation of amino acids and their polymerization into proteins, in the absence of a genetic apparatus the same conditions could be achieved in the next generation of vesicles. Let us imagine that a certain vesicle had produced some really useful protein which favoured the viability of that vesicle. How would such a vesicle create other vesicles which could also create this protein? That is, in this helter-skelter of macromolecular synthesis, how would there arise the stability necessary for guiding this synthesis into producing another useful protein?

One possibility is that the vesicular system and the genetic system arose separately and concurrently. At some point the two systems fused. This would mean that a major step leading to the origin of life resulted from a dramatic combination of separately evolved systems.

Clay Minerals and the Origin of Life

In reviewing what has become known as the 'clay hypothesis', Hartman (1986, pp. 10–12) points out that all contemporary life forms involve, first, nucleic acids, which are the chemical basis of replication and mutation, and second, proteins, which mediate the general metabolism of the cell. This represents the dichotomy between genotype and phenotype, and is much too sophisticated a system to have arisen all at once. Furthermore, the search for simpler, more primitive, systems must go beyond nucleic acids and proteins. The clay hypothesis posits that inorganic crystals would have had a major role to play in such pre-biotic systems (Cairns-Smith, 1986).

Clays are ancient and ubiquitous. They are found in meteorites (Chang and Bunch, 1986), and about half of all sedimentary rocks on Earth are made of clays. Duplication, differentiation and recombination characterize the processes by which clay minerals arise. They are the product of both degradation and synthesis. Weathering of rocks involves not only grinding rocks into small bits, but also the dissolving of various minerals in these rocks and their crystallization from supersaturated solutions. Clay crystals build up through polycondensation. They are continuously generated from very dilute, yet supersaturated, solutions – the product of the weathering of materials such as feldspars, which in turn are continuously generated through geological processes.

Clays, because of their enormous surface area, are ideal for filtering, for chromatographic processing and for catalysis. This means they can act like the primitive equivalent of proteins, organizing 'metabolic' events: for example, Mortland (1984) has described such a system, composed of a Cu^{2+}-smectite and the well-known co-enzyme pyridoxal phosphate. This 'pseudo-enzyme' has been shown to deaminate DL-glutamic acid and glutamine, but not the closely related aspartic acid and asparagine. As Pinnavaia and Mortland (1986) point out: 'The deamination of glutamic acid is not a particularly interesting reaction in itself. But the work demonstrates the activity of the silicate structure in the catalytic process and suggests other important reactions might be feasible such as transamination and decarboxylation'. As such, it 'is relevant to the possible role of clay–transition metal cation complexes in pre-biotic chemistry and chemical evolution'.

These and other observations indicate that clay mineral–protein complexes are able to perform the 'phenotypic' functions of biological systems if, and this is a big if, they were properly organized and coordinated. That is, a great deal more information would need to be added to make such pre-biotic, semi-organic systems work like living cells. Part of such an information requirement could be met by the availability of a 'genotype' master system, a system which may be stably reproduced, yet which exhibits the possibility for variation.

Before a genetic apparatus could be harnessed to act as both master regulator of the phenotypic systems and perpetuator of the information which assured the survival and reproduction of such complexes, there had to exist some stable information reproducer in the first place. This leads to the concept of the 'naked gene', which Mackay (1986, p. 142) defines as a system in which: 'A message is simply reproduced'. A naked gene, originally, need not relate to anything. Mackay also showed that, basically, a crystal matrix has the characteristics of an abacus on which it is possible to write an arbitrary message. Clay crystals, like abacuses, are combinatorial structures with a large number of metastable states on which it is possible to write a large number of arbitrary messages.

We are used to messages being carried on *uni*dimensional carriers. The words on this printed page are strung along in one direction, like beads. This is equally true for data stored on a computer tape, or music on an audiotape. It is still true when compressing data, or music, onto a disk – the information remains in linear form even if it is now in the form of a coil. Likewise, DNA and RNA are unidirectional message carriers. As Mackay points out, both 'human language and the genetic code, depend on messages written and read in a linear sequence'.

Crystals, on the other hand, can be two-dimensional or three-dimensional message carriers. Two dimensions makes it easier to reproduce the information. However, it is more difficult to access and edit information unless the two-dimensional matrix is broken up into easily accessible rows and columns. Three-dimensional information systems become more difficult still, unless, like in a book, they are made up of easily separable leaves.

To get a feeling for crystal genes, envisage a brick floor made up of yellow and red bricks. If the nature of the crystal growth is such that normally only a yellow brick stacks on top of a yellow brick, and a red one on top of a red brick, then the two dimensional pattern would be reproduced faithfully in the third (vertical) direction. Assume now that the nature of the inter-brick bonding is such that cleavage of the whole crystal can occur only in the horizontal plane, i.e. in a plane perpendicular to the axis of growth. Once fractured, the bottom half can continue growing upward, as before, while the top half can grow downwards – both perpetuating the original yellow/red pattern. (Actually, the two surfaces, top and bottom, would not look identical but would represent mirror images of each other.) Any stable irregularity (such as substituting a red brick with a yellow brick, or even a blue one, would be efficiently transmitted from one generation to the next as long as growth was only in one direction (vertical) and cleavage occurred neatly only in a plane perpendicular to axis of growth (i.e. horizontally). In such a system, the information being replicated would be a two-dimensional pattern. The information carrier, however, would have a robust, three-dimensional structure, with the information carried within it highly redundant. Such a system may be nowhere near as sophisticated and efficient as the enormous amount of information perpetuated in living cells, but in the pre-biotic world of a few thousand million years ago robustness was probably more important than elegance.

With the above as background, Cairns-Smith (1986, pp. 143–52) has considered a number of systems which would provide a class of objects which could act as 'proper crystal genes'. He concluded that there are four general requirements that are critical: (1) *disorder*: to provide information capacity; (2) *order*: for replicative fidelity; (3) *growth*: to duplicate the information; and (4) *cleavage*: to complete the replication process.

As one of the pioneers of the *clay hypothesis*, Cairns-Smith has raised some profound questions about the origin of life and provided thoughtful answers and useful analyses, such as the one above. Of the four general requirements, however, one needs to be cautious about how one interprets the first: *disorder* does not provide information! Mackay's naked gene needs no disorder to exist and multiply – only order, growth and cleavage. Disordering it would in fact destroy it. Information is a function of organization: information is *inversely* related to disorder and the entropy which measures this state. However, disorder can provide the mechanism for altering the structure of a system so as to allow a 'mutation' to appear. If there existed no possibility for introducing *variations*, then there would be no possibility for the system to evolve. Given that such variations in crystal systems may involve specific individual loci, where, for example, one atom or atomic grouping is replaced by another, it might be conceptually more helpful to restate Cairns-Smith's first requirement as: '*disorder*, to provide *mutations*'. That is, the term 'information capacity', derived from the communications engineers, should be replaced by the more relevant term 'mutation', used by biologists. It would also be more lucid if this, the first of the four requirements, were made the last.

To sum up, small molecular units such as silicic acid and hydrated metal ions can combine in a multitude of ways and can give rise to a large number of stable combinations. Clay crystals, therefore, may be classed as aperiodic crystals whose crystalline character provides both regularity and stability. Clay crystals were plentiful in pre-biotic times. As such, they could have performed, even if only in a much less efficient manner, functions currently carried out by nucleic acids. Other clay minerals, in an analogous fashion, and again with less efficiency, would have been able to carry out the catalytic functions currently carried out by proteins. Clay pseudo-genes and clay pseudo-enzymes, coupled to a mixture of organic compounds in an aqueous medium, would have set the stage for numerous interdependent systems of advancing complexity.

In this manner, *hierarchies of complexity* might have emerged, reaching levels of thermodynamic improbability impossible to describe with mere astronomic numbers. However, in each case the *feedback loop* remains the basis for stabilizing the integration of complex subsystems. For example, in contemporary living systems no nucleic acid synthesis can occur if the right amount and type of protein enzymes are not available. Conversely, no protein synthesis takes place if there are no messages emanating from nucleic acids. There can be no advanced levels of dynamic organization without an extensive network of feedback loops.

Is the Clay Hypothesis Enough?

It is the requirement for feedback loops which limits the probability that clay pseudo-genes and pseudo-enzymes could, by themselves, evolve into living organisms. If the information that is stored in the clay pseudo-genes is transferred to the surrounding medium, that surrounding medium most not be so large that the message becomes too diluted and its receptor too far removed for the system to function. Only three scenarios seem possible to make the clay system work.

The first is that the Earth is covered with large patches of primordial slime in which both the clay pseudo-genes and pseudo-enzymes operate in a concentrated mix of organic compounds and minerals. The entire sea of primordial slime would behave as a single proto-biotic system.

The second scenario is restricted to pools sufficiently small to achieve the critical mass concentration which allows the message sent out by the clay pseudo-gene to reach its receptor, and conversely that the receptor system is able to involve molecules in sufficient proximity to each other so as to create more pseudo-genes.

The third scenario is that the clay genes and catalyst become 'entrapped' inside a vesicle, as envisaged by Morowitz, to form the kind of corpuscle defined by Teilhard de Chardin.

All three scenarios are possible, but the first two are less probable. Whole seas filled with primeval slime seem unlikely because their residue should leave behind a trace for geologists to discover. As Yockey (1995) has repeatedly pointed out, such evidence has not been uncovered.

Small pools might be too rare and too scattered to leave behind a geological record that might have been found. The trouble with this scenario is that small pools dry up or are flushed out by springs and other floods. Therefore, small pools would provide such an unstable environment as to make it difficult for a proto-biotic system to flower into a living organism.

The best bet for small pools would be intertidal rock pools, high enough to be replenished only occasionally, e.g. during a spring tide, and deep enough that such pools would not dry out. These high, intertidal pools are likely to have an increased mineral content, supplied by the rocks themselves, by the sea-water and by volcanic 'rain'. Concentrated by evaporation and exposed to atmospheric gases and ultraviolet radiation, such pools could constitute cauldrons of pre-biotic activity.

However, the pool might still be too large to allow the chemical feedback loops required for the orderly reproduction of chemical events. The matter becomes more credible by invoking the micro-environment created by vesicles. By placing the clay pseudo-genes and the clay pseudo-enzymes inside a membrane, we set the stage for the kind of information feedback which might allow order to emerge from random chemical events. Thus, both mechanisms, those invoked by the clay hypothesis and those described by the vesicular hypothesis, may well have created the first major symbiotic interaction to play a part in the origin of living systems.

The next major step after the creation of a single successful vesicle capable of producing, if inefficiently, more pseudo-genes, would be to have such a vesicle reproduce. Since the membrane structure is such that it is easy to simply add more to it, the vesicle would balloon out. If sections of the membrane were to interact with other compounds so as to result in polar sections of the membrane, such portions might become stuck to polar structures inside the vesicle or with each other, while the rest of the membrane continued to balloon out. This could become a mechanism for the vesicle to pinch off new portions, budding, as it were, like some yeast cells.

Modern cells have evolved several complex mechanisms for pinching off vesicles from existing cellular membrane systems (Rothman and Orci, 1996). In pre-biotic times, this much less sophisticated pinching off would not produce two vesicles of the same size, nor necessarily of the same composition. As such, this process may be referred to as ***pseudo-reproduction***, in which the 'offspring' would not necessarily be composed of the same material, nor carry the same information as does the parent vesicle. Many times such offspring would lack one or more essential elements and therefore would become a non-viable dead end. On the other hand, this process might occasionally have led to daughter vesicles whose content had a higher survival value. It is at this point that one would observe two new features: first, the vesicular system is beginning to cross the threshold dividing proto-intelligent from intelligent systems; and second, the system would henceforth begin to evolve as if it were a true biosystem: traditional Darwinian evolution would come into play.

Assuming that the above scenario has validity, one must then rely for further progress not only on the accidental emergence of new features, such as the creation of chromophores, but also on the repeated combination and recombination of separately evolved systems. By these processes, proto-biotic systems would evolve from possessing relatively inefficient mineral catalysts (the pseudo-enzymes) to highly efficient protein catalysts (the cell's enzymes), and from clay pseudo-genes to the nucleotide-based genetic apparatus of all extant organisms.

Before DNA

It is now generally agreed that before DNA emerged as the prime repository of the genetic code, there existed an 'RNA world'. RNA, as demonstrated by retroviruses, as well as by the myriad of ***information functions*** it carries out in the modern cell, is capable, at least in theory, of organizing and maintaining life forms. What occurred prior to the establishment of an RNA world is a matter of debate.

De Duve (1995) has suggested that a disorganized but robust 'protometabolism' based on a 'thioester world' could have generated both the mononucleotides and the polynucleotides which, ultimately, would have provided the RNA templates for self-organization. The thioesters theory is strengthened by two observations: first, hydrogen sulphide (H_2S), the rotten egg odour gas emitted as a result of volcanic activity, would provide the reducing power necessary for creating such thioesters during pre-biotic times; and second, the importance of that evolutionary origin would be reflected in present-day metabolism – certainly, thioesters are crucial in a number of contemporary metabolic reactions.

In my opinion, de Duve's postulation has much to commend it. It will probably prove to be correct, but only in part. The idea that thioester metabolism was well established before the RNA world came into being makes good sense. However, to extend this concept, that thioesters were the primary route to biogenesis – that a multiplicity of thioester reactions finally sorted themselves out to create life – does not ring quite true. The scepticism expressed here is the same as that applied to Morowitz's vesicle theory: how in the helter-skelter of numerous reactions does a system achieve the stability to reproduce itself generation after generation? De Duve's suggestion that the environment restricted all reactions to just a few, stable, self-reproducing ones must, at present, be considered as improbable.

This brings up another point: de Duve – and his view is shared by other biogenetic theorists – dismisses the clay hypothesis on the grounds that vestiges of a pre-biotic metabolism ought to be evident in today's metabolism (congruence). There are two reasons for faulting this line of argument: first, such vestiges do exist; second, 'if you don't know where you started from you wouldn't know how to get there'.

The first assertion rests on the fact that ***inorganic chemicals*** are as important as thioesters in metabolism. Trace elements such as iron, zinc, copper, cobalt and manganese, which play such critical roles in enzyme catalysis, probably reflect a pre-biotic metabolism. It would seem much more likely that the organic part, the protein moiety, would be added, bit by increasingly efficient bit, onto the metal elements, rather than that some complex but ineffective apoenzyme would suddenly discover the metal catalyst it needed to become active. The element molybdenum, which is quite abundant in natural waters, is strange (Stiefel, 1996). Molybdenum enzymes fall into two categories: the nitrogenases, responsible for biological nitrogen fixation, and other enzymes which embed the molybdenum in a common non-protein organic molecule called molybdopterin. In thermophilic microorganisms this same molecule appears, except that tungsten is substituted for molybdenum. The existence of molybdopterin supports the idea that the earliest catalysts did not involve proteins as a major component. A similar argument can be made for porphyrins surrounding an atom of iron and other elements.

The second assertion simply assumes that, by the time chemical evolution was approaching biological evolution, most of the modern systems were in place. The ecology of preCambrian oceans bears little resemblance to the ecology of a coral reef. Likewise, one would not guess by looking at a frog that, in its youth, it used to be a kind of a fish, or that a butterfly was a kind of worm. For that matter, one would not guess that a human being, at its start, was a tiny 'egg', too small to be seen except with a microscope.

In our present state of knowledge, it is a simple matter to trace a single cell into a full-grown human being, a caterpillar into a butterfly, a tadpole into a frog, and the evolution of marine ecosystems from preCambrian times to a coral reef. But, in each case, we know the beginning point and something about how these systems manage to metamorphose. At the moment, the same cannot be said for the pre-biotic phase of our planet's history.

To illustrate this point: the general assumption (also in this chapter) considers that, very early on, there was a need for the appearance of chromophores to convert the light energy of the Sun into metabolic energy of the cell. This is probably wrong. Recently, two researchers uncovered what appears to be the first ***subsurface lithoautotrophic microbial ecosystem***, existing a mile below the surface of the Earth (Stevens and McKinley, 1995). These lithoautotrophic bacteria obtain their energy from hydrogen released when the ferrous silicates of basalt rock react with water. By utilizing this hydrogen to combine with the carbon dioxide dissolved in water, the bacteria are able to synthesize all the carbon compounds needed for life.

If inorganic hydrogen generation was commonplace during the early stages of life formation on Earth, then this kind of lithoautotrophic metabolism would have arisen first and would have presented, one might say on a silver platter, the whole constellation of enzymes needed to convert carbon dioxide into organic carbon. All that the chromophores and associated photosynthetic apparatus would be adding (and its addition would be highly significant) would be a new source of hydrogen by the photolysis of water.

The reason why the clay hypothesis should not be dismissed lightly at this point in history is that the concept of the naked gene, as defined by Mackay, provides a starting point. True, such a gene merely reproduces a message, a message which has no meaning for any metabolic system. However, it does have the potential for faithful reproduction of information. A parallel situation must have prevailed for the chemical reactions taking place in pre-biotic systems. Such chemical reactions would have been totally uncoordinated and unregulated, creating their products with no regard to the significance for any future system which might reproduce itself. The same may also be said for the formation of the early membranes – enclosing nothing of any consequence.

However, the combination of the naked gene, placed in an active metabolism, including de Duve's thioester world, further energized by the presence of light-transducing chromophores and shielded from the vagaries of an indifferent environment by being encapsulated in Morowitz's vesicles, would mean that all the elements for creating a proto-cell were present. Inefficient as all these elements might have been, the possibility would arise that the whole system might begin to function as an organized unit.

It is here – the development of feedback loops which regulate the flow of information among the many parts comprising such a proto-cell – that all theories of the origin of life flounder. A new factor must have come into play at some point to organize the whole into a corpuscle. What that primordial event was, which caused chemical evolution to become superseded by biological evolution, has not yet been discovered and remains one of the last great mysteries of the biological sciences.

Once the proto-cell has become sufficiently organized to cross over the divide into a true life form, all systems would become more efficient. Polymers would evolve to become more useful to the architecture of a cell's structure. The primitive membranes of the early vesicles would become better at screening out undesirable materials, admitting growth factors and eliminating waste products. They would also evolve into the immensely complicated membrane systems of advanced cell types. Inorganic catalysts would be displaced by more efficient organic ones, which would ultimately evolve into protein enzymes and RNA ribozymes. In the meantime, genes of clay would combine with organic molecules until finally they would be displaced by the much more efficient RNA and still more reliable DNA systems.

All of this is still a matter of speculation. At this point in history it should not be taken too seriously. De Duve's book *Vital Dust* provides a highly intelligent and readable account of what might have happened. Furthermore, the professionalization of the field of biogenesis was given a real boost when the US National Aeronautics and Space Agency (NASA) created an 'exobiology' group. The rationale for spending nearly a million dollars per year to support such a group is that the search for life on other planets would be greatly facilitated if more was known about how life originated on this planet (Cohen, 1995). Thus any definitive discussion of the origin of life must be postponed to some future time.

What is important to our discussion is that all these pre-biotic systems – membranes, thioesters and other metabolites, chromophores, and naked genes – all possess information and all are capable of engaging in information processing. All systems may have existed in their own right. None, originally, had anything to do with developing into a cell – that came later. Only with the advent of feedback loops, coordinating the interactions between the various components, did the organization of independent molecular systems into coherent living matter become probable.

When one looks at the metabolic activity of a modern cell, it would *appear* as if the multitude of enzymes, with their various feedback loops and other control machinery, are engaged in signalling and in information processing. This resemblance is much more than superficial: the main task of enzymes is to create or destroy chemical bonds. In so far as the organization of matter, and hence its information content, is wholly dependent on the bonding which occurs between various sub-units, the creation and destruction of bonds is directly linked to changes in the information status of the system.

The difference between a ***vesicle*** and a ***corpuscle*** resides in the fact that the latter has developed feedback loops. The first step in creating such a feedback loop is when the output of one reaction becomes the input of another. If the second reaction produces an output which becomes an input into the first, the loop is complete. However, a more likely scenario is one in which a series of reactions forms a chain in which each preceding reaction yields an output which serves as an input for the next reaction. Ultimately, a chain of such interactions could turn back on itself to create a cycle. Such metabolic cycles are commonplace in today's cell metabolism. In each case, it must be emphasized, such a metabolic cycle requires an input of energy to drive it.

The idea that the output of one system serves as an input into a second system is a most important concept. It sets the stage for the interdependence of various systems. If a feedback loop is established along the lines discussed above, a new level of complexity is achieved. This in turn would constitute a step towards a higher level of information processing – that is, it would produce the hierarchy of information-processing systems we see around us.

DNA as Part of an Information-Processing System

Lila Gatlin's book *Information Theory and the Living System* (Gatlin, 1972) begins with a most useful definition of life: 'Life may be defined operationally as an information processing system – a structural hierarchy of functioning units – that has acquired through evolution the ability to store and process the *information* necessary for its own accurate reproduction'. The mechanism by which a cell achieves such accurate reproduction is, of course, DNA. The great feat of DNA is the ease with which it can duplicate itself. When split longitudinally, each half can act as a template, regenerating the other half. In this respect, it acts like the hypothetical clay gene. Unlike the clay gene, however, DNA is able to act not only as a vehicle for transmitting information from one generation to the next (its *genotypic* function), but also is able to act as the regulator of the cell's metabolism (its *phenotypic* function), including its own synthesis.

One reason for DNA's success must reside in the fact that its information has the property of being read easily by biosystems.

The cell may be viewed as a computer which is reading its DNA as the program. It differs from electronic computers in that information processing carried out by electronic computers manipulates ***abstract*** information (e.g. the human language), whereas the information processing carried out by cells involves ***physical*** information: the rearrangement of atoms and molecules. The cell is able to experiment by allowing variations in its genetic material to occur. Successful experiments survive and reproduce, having transcribed the new variation in their DNA (from whence it originated). Unsuccessful programs are eliminated. Thus there has evolved a mechanism for transgenerational learning, and because all extant biosystems show this property they must all be classed as intelligent.

Intelligence and the Control of the Environment

To conclude this chapter on bio-intelligence we need to introduce one more concept. It is a concept which also applies to the previous two chapters on proto-intelligence and the spectrum of intelligence.

As one examines the spectrum of intelligence, from systems which appear to show no intelligence to human global society, there emerges a new principle: the relationship between a system and its environment. At one extreme is a cube of sugar dissolving in a cup of tea. Although highly organized, the cube is totally controlled by environmental events and exhibits zero intelligence. At the other extreme is a technologically advanced society deciding to divert the waters of a river so as to irrigate a plain to provide an assured food supply for the entire population.

In considering the limits of the spectrum of known intelligence, as well as the various stages in between, there emerges, therefore, a new principle: the intelligence exhibited by a system, may, at least in theory, be measured as a ratio, or *quotient*, of the ability of a system to *control* its environment versus the tendency of the system to *be controlled* by the environment.

Using the environment to best advantage is one way of exercising control over the environment. A system *adapting* to a stable environment may be considered as having developed a form of partial control: it is a way of both exploiting the environment for the benefit of the system and reducing the likelihood of the system being destroyed by its environment.

The key concept, therefore, is this: *intelligence is a measure of the system's ability to respond effectively to changes in the environment, thereby enhancing its survivability or reproducibility.*

In biosystems, this involved, initially, various forms of adaptation which allowed systems to develop internal homoeostatic mechanisms to offset changes in the external environment. Later, higher forms of intelligence were able to influence their environment directly and control such changes. Ultimately, such systems began to restructure their environment.

Restructuring of the environment is exemplified by ant-hills, birds' nests and beaver dams. However, the epitome of this process of environmental control has been achieved by contemporary human society: the reader, looking up from this text, is likely to see very little that has not been created by human hands. Even outdoors, the trees and meadows reflect human intervention. Only a virgin forest, a desert, the mountains, the oceans, and the clouds and the sky have been left relatively untouched.

The evolution of human civilization may be viewed as a process which results in an ever-increasing control over its environment. One must be clear that the environment of any given society includes not only its physical environment (soil, rainfall, natural resources etc.), but also its biological environment (plants and animals, pests and diseases) and its social environment (neighbours, trading partners, enemies). For example, a society may have assured itself of a steady food supply only to fall victim to a plague or be decimated by war-like neighbours. The appearance, development and demise of a wide range of cultural institutions during the evolution of human societies represent efforts on the part of those societies to maximize control over their total environment.

From termite mounds to human cities, *life*, utilizing *intelligence* as its tool, has been striving to create surroundings of its own choosing, thereby reducing the uncertainties of an indifferent environment. A measure of the effectiveness of intelligence, and hence of intelligence itself, is the success with which the system is able to create its own favourable environment.

Summary and Conclusion

How life originated remains one of the great mysteries of the biological sciences. However, what should become apparent from our cursory discussion of the matter is that a variety of mechanisms must have been brought into play in order to achieve the evolution of self-organizing chemical systems into self-reproducing biological systems. Furthermore, such a process makes it likely that the shift to self-reproducing biosystems may well have occurred more than once.

The fact that all known life forms existing on this planet have the same genetic code implies that we are all descended from a single primordial cell – a cell whose superior organization and information processing capability overwhelmed all competition. Whether such competitors ever existed is hard to prove, although there remains the possibility that traces of them have survived in the deep recesses of our Earth and that the exploration of deep ocean ecosystems, particularly in the area of deep ocean volcanism, may yet uncover them.

The origin of life, as a phenomenon, takes on new meaning when one considers that what is happening among us today is a similar process of proto-intelligence evolving into true intelligence. It is this process which will be the subject of our next chapter.

· 7 ·

Machine and Electronic Intelligence

Mechanical Intelligence

In Chapter 3 we alluded to both the flyball governor on a steam engine and to the thermostat in a house as examples of mechanical information-processing systems. Here we wish to examine another transition, from information-processing systems that are unintelligent, or which exhibit mere proto-intelligence, to those which might be considered intelligent.

For example, in line with the definition that a goal-oriented system exhibits intelligence if it analyses its environment, and on the basis of that analysis engages in a response which results in the achievement of that goal, a thermostat certainly exhibits intelligence – even if it is of a very low order.

The thermostat is a part of a larger system – a house. As a system, a house as a whole, equipped with a heating system controlled by a thermostat, can exhibit a somewhat higher level of intelligence. Many people living in Canada and in the north of the USA spend a few weeks or months escaping the harsh winter by going to warmer parts of the USA, Mexico or the Caribbean. In doing so, they often leave behind – unattended – their houses to face the cold. In some instances, the escaping owners may drain all the pipes and shut the whole system down completely. Frequently, however, if the home is filled with plants, a piano or other items which might be damaged by excessive cold, the thermostat is merely turned down somewhat. The house is now left to its own devices to face the cold without incurring damage to itself or its contents. If the heating system is considered to be an integral part of the house, in the way that its plumbing, wiring, insulation etc. are a part, then the house, by avoiding the calamity of frozen and cracked pipes, is enhancing its own survival by using its thermostat to both analyse the environment and, on the basis of that analysis, initiate an appropriate (intelligent) response: activating the heating system.

This is the basis for saying that a house may be considered to be an information-processing system that exhibits the property of intelligence.

Can a House Learn?

The above is an example of the difficulty of drawing the line between intelligent and non-intelligent, or proto-intelligent, behaviour. One could argue that the above is only a proto-intelligent system because of its inability to ***learn***. To get around this criticism, and devise a mechanical IPS which has the capacity to learn, one would need to add a memory and a flip-flop circuit. A flip-flop circuit allows an electric current to flow through one of two channels, but if it flows through one, it cannot flow through the other. This concept of the 'flip-flop circuit' is extremely important and is worthy of deeper consideration. Coupled to a memory device, a simple flip-flop circuit provides us with the rudiments of a machine system which is capable of learning.

The nature and implications of such circuitry may be better understood by engaging in a 'thought experiment'. Consider two rooms in a house whose heating is determined independently because each of the two heating zones is controlled by its own thermostat. Many modern houses have such a system.

Now let us introduce a novel feature into the system: what would happen if, when a thermostat, in response to cool temperature, not only turns on the heat in its own room, but also, as it closes the switch, activates a small electric heater which has been installed right *below* the other thermostat in the other room. This electric heater is much too small to heat the room, but it heats the thermostat sufficiently so that the thermostat will not turn on. That is, the second thermostat is 'fooled' into sensing that the room is warm, even though it may be quite cold. Now let us reciprocate. If the second thermostat turns on, it also activates a small heater next to the first thermostat. What this means is that whichever thermostat turns on first will prevent the other one from turning on.

Let us refer to the rooms as room A and room B. If we turn on A first, B is suppressed, because even though it may be cold in room B, the small thermostat heater keeps B's thermostat from closing. Room A is getting warmer, while room B keeps getting colder.

Here is one way in which the situation could change: room A heats up enough to shut off its own thermostat; as A's thermostat shuts off, it also shuts off the small electric heater next to thermostat B, which, the room being cold, now activates B. Once B is turned on, it suppresses A because it causes the small electric heater blowing on the A thermostat to be activated.

The above represents a typical flip-flop system. As such, however, the system is not yet capable of *learning*. Suppose we prefer one of the two rooms to be warmer than the other. To get the system to perceive this pattern and remember it, we must introduce a *memory* device. This could be done by introducing a *counter* such that every time a thermostat turns the system on, it registers as one turn on the counter. In other words, every time a thermostat turns on, it scores a 'point'. One could then couple this counter to the thermostat such that for each 100 points, the thermostat setting is lowered by one degree. So, for example, if the thermostat had originally been set to turn the system on at 20 °C, after turning on 100 times it would be set for 19 °C.

To make the system still more responsive, we should introduce a second mechanism which deducts one point from the other thermostat every time the first thermostat gains a point. This means that if thermostat A earns 100 points (lowering its temperature setting from 20 °C to 19 °C), thermostat B would lose 100 points, raising its setting from 20 °C to 21 °C (assuming it too started with a setting of 20 °C). This means that if the temperature falls to anywhere near 20 °C, the thermostat in A would turn on first (it is now set for 19 °C) and, at the same time, would suppress the thermostat in B.[1]

The introduction of the above mechanisms into the thermostat information-processing system establishes a general rule: the more one part of the system is used, the more likely it is to be activated in the future. This is an enormously important principle when considering the functioning of the neural network which comprises the human brain; and, of course, it is the principle on which neural network computers work.

Finally, if we walk into a room that is too cold for us, there should be a manual override system such that we can activate the thermostat by pressing a switch. Furthermore, every time we close a switch ourselves the thermostat earns a point, while the thermostat in the other room loses one. This means that every time we come into a room, find it too cold and adjust the thermostat, the thermostat system begins to recognize which

1 Obviously, one would want to set an upper limit to the number of points earned by a thermostat, otherwise its setting would end up so low that it would never shut off. Under those circumstances, if A were the favourite thermostat, room A would be heated continuously, while room B would remain perpetually cold.

room we are more likely to enter and shifts its preference as to which room should be kept warmer.

Thus, by coupling several proto-intelligent systems we have devised a system which is able to learn, and hence must be credited with a measure of true intelligence because it:

1. *analyses* its environment (by measuring the temperature of the room);
2. *responds* to it (by closing or opening the thermostat switch, thereby calling for or withholding heat);
3. *detects a pattern in its own behaviour* (by counting how often the thermostat switch closes in either room);
4. *remembers* that pattern (by recording the number of points scored by each thermostat); then
5. *favours the successful pattern*, so that one set of behaviour patterns (e.g. heating room A) increasingly dominates over the alternative set (heating room B); and
6. in doing so, exhibits *learning* behaviour.

A final bit of sophistication would be not to require any manual manipulation of any of the thermostats, but simply to have a sensor which counts the number of times a human being enters the room, and for how long.[2]

---

2 We could retrain the system by using room B more often, or by deliberately overriding and activating thermostat B. Every time we cause B to close, it earns a point while A loses one. If it takes 100 points per degree, and the previous history of the system was such that A ended up at a setting 10 degrees lower than B, then it would take 500 manual interventions to earn B 500 points (worth 5 degrees) while A loses 500 points, thereby setting both thermostats to the same temperature. Further manual overrides favouring B would give B an advantage. From here on in, B would predominate – warming room B and keeping room A cooler.

Note that, given the above assumptions, it would require over 500 human interventions to teach the system that it should reverse its previous stance. The system would be a slow learner. However, one could alter the system in at least two ways to make it a ***fast learner***: first, one could reduce the upper limit of the temperature differential. For example, if the difference between B and D were never allowed to exceed 5 °C (instead of 10 °C), then it would take only half the number of interventions. Second, if, instead of each human intervention resulting in a one point differential, it resulted in 5 points, or even 10 points, then the difference between the two would be reduced 5 or 10 times as fast – the system would now learn the new pattern much faster.

One could manipulate the system in many other ways: for example, one could put different size heaters next to the thermostats – the smaller the heater, the more important the room temperature becomes. Such a system would mean that a smaller heater

In biological systems, in particular in the mammalian brain, the equivalents of all of the above systems are probably in operation. For example, brain structures controlling various body movements have pronounced architectures so as to give a bias to activating and coordinating muscles crucial to the survival of the species in a most efficient manner. The proper activation and coordination of these muscles must still be learned, but animals which climb trees (e.g. monkeys) or swim in the ocean (e.g. dolphins) differ not only in gross anatomical features but also in their brain structures. In humans, for example, a large portion of the brain is dedicated to coordinating the movement of the hands and fingers.

In the human brain there are mechanisms for learning things rapidly and for *unlearning* things rapidly. There also exist mechanisms for learning things rapidly but then unlearning them only very, very slowly. This last is typical of the 'imprinting' phenomenon, which we discussed earlier: some things can only be learned by an immature animal during the early stages of its life cycle; after that the network appears 'frozen', so that adults cannot learn or relearn. In humans, learning to speak a foreign language without any trace of accent is such an example.

The important point in all this is that it is possible to begin with the simple negative feedback loop of a home thermostat system, expand this to the double negative feedback loop of a flip-flop circuit, then add memory devices, such as counters, which are also tied into the system by feedback loops, and end up with a system which is capable of learning.

Note that in all this we used a flip-flop system involving only two thermostats and two rooms, with only two possible pathways. Either room A is heated or room B is. Extend this now to many rooms in the house, and consider the possibility of many different patterns, not just two, and we begin to see the outline of a rudimentary neural network – even though it is a network dedicated merely to heating the various parts of a house.

The above thermostat model involves a machine intelligence based on old-fashioned mechanical and electro-mechanical devices. Think how much more powerful and sophisticated could be systems using the fast-

in room A would favour thermostat A over B. Alternatively, one could set the temperature setting of A lower at the start, so that even though the temperatures in the dining room and bedroom might be the same, A would be activated before B. One could vary the size of the radiators, or vary the insulation or introduce timers which interrupt the thermostats, all of which would vary the time for which one circuit predominates over the other. Finally, one could introduce delays into the system so that, for example, B could not switch on until one hour has elapsed after A has shut off. All of these mechanisms would create a system with a strong bias, in this case, favouring heating room A (by activating thermostat A) at the expense of room B.

switching solid state technology available to computers. Furthermore, instead of only a single intermediate layer, one could have hierarchies of layers, as in the human brain, relating the input to numerous existing patterns, weighting it at each layer and refining it so that the output 'makes sense' within a given context of prior inputs and associated memories.

Deep Thought and Deep Blue

Chess-playing machines, on occasion, grab headlines. For instance, in early 1996 the headlines trumpeted the match between Gary Kasparov, considered the world's greatest chess player, and ***Deep Blue***, the world's most powerful chess-playing computer. Kasparov won the match, but he was shaken up.

The earlier match between Kasparov and ***Deep Thought***, IBM's predecessor to Deep Blue, has been reviewed briefly in an earlier work: *Beyond Information* (Stonier, 1992). It might be instructive to reemphasize certain thoughts which came out of that match. Feeling rather contemptuous of ***Deep Thought***, Kasparov was quoted (Hsu *et al.*, 1990) as saying that his creativity and imagination must surely triumph over mere silicon and wire.

The creators of *Deep Thought*, Hsu, Anantharaman, Campbell and Nowatzyk, countered Kasparov with the observation that when the two met again, it should not be viewed so much in terms of man against the machine; rather it should be considered as 'the ingenuity of one supremely talented individual' pitted against 'the work of generations of mathematicians, computer scientists and engineers'. They believed that the outcome would determine whether 'collective human effort can outshine the best achievements of the ablest human being'.

Kasparov's match against ***Deep Blue***, in February 1996, produced a profound change in his attitude. It was not merely that ***Deep Blue*** beat him in the first game of the series, but that the machine – mere silicon and wires – gave him the feeling, at times, that it played with insight, anticipating not only its own, sometimes very original moves, but anticipating Kasparov's as well.

What Kasparov was observing, and had not taken into account earlier, was the power of evolution, which applies as much to technological information-processing systems as it does to biological information-processing systems. ***Deep Blue*** evolved from ***Deep Thought***, which had evolved from its earlier predecessor, ***ChipTest*** (Horgan, 1996).

As more and more information-processing capacity was added, levels of complexity increased. It is not clear yet when such a system would undergo the equivalent of a phase shift. This is like water being heated: the temperature rises slowly with not much happening, and then suddenly it boils over. Kasparov seemed to intuit that ***Deep Blue*** may have achieved such a phase shift: quantity had become quality. As one reporter, William Dowell (1996), observed: when a computer can calculate so fast, and look so far ahead in the game, it rises to another level. The matter parallels what happened in biological evolution: 'There are neurons firing in lemmings and squid, but put them together in gigantic enough numbers and fantastic enough array, as in humans – and behold, a thought, popping up like a cartoon bubble from the brain'.

By May 1997, an upgraded Kasparov, training for the rematch, had regained his confidence. A champion who had never been beaten by anyone or anything was ready to show this crystal-brained piece of hardware who was boss. ***Deep Blue***, however, had also been upgraded. Not only was it programmed to play a better opening and a superior endgame, its capacity to examine board positions had been doubled from one hundred million per second to two hundred million.

Kasparov won the first game, *Deep Blue* the second. The next two were draws. In the fifth game Kasparov had the advantage of playing with the white pieces. He appeared to be winning until *Deep Blue* pursued a brilliant endgame, forcing a draw. The effect was devastating. After five games, Kasparov had not been able to demonstrate his superiority. Demoralization set in. He simply blundered into defeat in the sixth (and last) game, thereby losing the match.

The headline **Checkmate: Machine is chess king** summed up the popular perception. According to press reports, Kasparov accused IBM and its computer team of cheating, complained that he had not been given the opportunity to study *Deep Blue*'s earlier games, would play it at *real* chess and beat it... and so forth.

Probably Kasparov could win a rematch if he had a chance to study *Deep Blue*'s games and ascertain its weaknesses – probably its middlegame – but, and this is a big 'but', only if *Deep Blue* were not improved further.

Be that as it may, as things stand the computer's victory is a blow to those who consider chess as the ultimate in human intelligent behaviour. For them, Kasparov's defeat diminished both chess and humanity. This is a rather negative attitude: *Deep Blue* is a creation of the human mind as well. However, its creation involved human collective intelligence and human creativity.

Deep Blue is only a logic machine. Long ago we accepted that computers are better at certain kinds of mathematical skills: this is merely an extension of that kind of thinking power. Logical thinking is only a small part of ***human*** intelligence: we are good at perceiving patterns, making connections, conceiving strategy, imagining things that never existed etc. Logical thinking as a guide to action has a great danger associated with it: a faulty assumption, when followed by impeccable logic, can only lead to a faulty conclusion. In so far as most assumptions are based on guesses, using ***only*** logic for solving most real-life problems is likely to lead to major mistakes. Humanity, as yet, does not need to fear that its intelligence has been outpaced by an electronic one.

Cellular Automata Machines

Some semioticians have used cellular automata on a computer screen to mimic cellular information processing events (Bogh Andersen, 1995).

The first volume of this series, *Information and the Internal Structure of the Universe*, described briefly a computer game called 'Life' (Stonier, 1992, pp. 124–5). Created in the 1970s by John Horton Conway at Cambridge University, it is a sort of video kaleidoscope. The screen is divided up into an array of squares, each square representing a *cell.* (Other figures and configurations are also possible.) In the square configuration, each cell (square) is surrounded by eight neighbours (one on each side plus one at each corner). Each cell may exist in one of two states – 'on' or 'off'. Whether it is 'on' or 'off' is determined by the configuration of on/off states of its neighbouring cells. The player determines the rules as to which pattern of neighbours' on/off states cause a cell to switch states. The computer ticks over at a rate which may be determined by the player – let us say, ten seconds. That is, every ten seconds the computer changes the pattern on the screen. The new pattern is determined *automatically* by the rules imposed by the player. Without actual observation and significant experience, complex rules make it almost impossible to predict the outcome after only a few cycles. Simple rules are easier: for example, the rule that a cell turns 'on' if its neighbour to the left is 'on', but turns 'off' if its neighbour to the right is 'on', causes the light pattern to tend to move to the right.

Although cellular automata do require computers and computer screens (and therefore matter and energy) in order to become manifest, they basically represent *universes made up of pure information.* As such, they can mimic a wide series of natural phenomena – in particular, evolving systems. Evolving systems are characterized by changes in their organizational states – hence by changes in their information states. This is probably why cellular automata can describe natural systems and phenomena with greater facility than any other method. Each set of rules represents an algorithm determining the behaviour of the system. All local systems are governed by the same algorithm, but each local landscape has been determined by local conditions which, in turn, reflect the locale's previous history.

In the long run, cellular automata will prove to be an important tool in trying to understand the phenomenon of collective intelligence. For example, consider the workings of an ant colony discussed in Chapter 3. The behaviour of each ant is determined by its interaction with the other ants it comes in contact with. Its responses are determined by the rules programmed into its DNA by the species' evolutionary history. Depending on the signals it receives, it may follow another ant, regurgitate food, attack or engage in any number of other activities. It may also simply ignore another ant if it does not receive any signals to alter its own behaviour. This last instance is equivalent to not changing its 'state'.

Weaver ants probably perform no more than 50 distinct behavioural acts – the majority involving communication. This is very much more complex than the two acts (on/off) exhibited by the cellular automata described above. However, the principle that the behaviour of any given unit within a complex system is determined by the state of the unit's neighbours remains the same. A computer architecture which would be specifically designed to accommodate cellular automata would be a great aid in analysing complex, evolving, apparently chaotic systems.

The field of cellular automata was introduced around 1950 by Stan Ulam and John Von Neumann. It is not the purpose of this section to review its history. The reader is referred to Toffoli and Margolus (1987), who have described cellular automata as 'stylized, synthetic universes', and a *cellular automata machine* as a 'universe synthesizer'. Their pioneering work on such machines recognizes that such machines represent a general paradigm for parallel computation in the same way that Turing machines do for serial computation. For the kinds of problems described above, cellular automata machines (CAMs) are vastly more efficient than their Turing counterparts. For example, for certain scientific applications it may be necessary to run through 10^{13} cycles. On classical computers, even if one allotted only a microsecond per cycle, it would require 30 machines several years to complete the computation. In contrast, CAMs, by virtue of their ability to engage in massive parallel computing, are able to deal with this kind of computational problem so efficiently as to allow a performance at least several orders of magnitude greater. Toffoli and Margolus have spent considerable time and effort in designing and improving such devices, and applying them to a wide range of problems. Clearly, CAMs represent another major direction in the future evolution of machine intelligence.

Why Computers Will Become Smarter Than People

It should become apparent from the preceding sections that computers will become ever more powerful thinking machines. We must therefore consider not *whether* the intelligence of computer systems will exceed that of individual human beings, but *when* (Stonier, 1988).

There are several compelling reasons why one would expect, at some point in the not-too-distant future, computers to outpace humans in virtually every and all intellectual tasks.

The first, and perhaps theoretically the most important reason, derives from the fact that whereas we can extend human knowledge virtually indefinitely, we cannot physically expand the human brain. Our (human) thoughts are trapped inside our mortal skulls. No such limitations exist for electronic intelligence: we can apply all advances in knowledge to creating more advanced forms of electronic intelligence. We can create devices whose lifetime far exceeds ours – approaching infinity if we allow for repairs and the indefinite transfer of data and processing capability. Furthermore, we can combine all kinds of computers to create supercomputers which ultimately allow us to place, into a single system, all human knowledge plus most (and given enough time, all) human neurological capabilities.

At the moment, computers are '*idiots savant*'. They show aspects of human intelligence such as the ability to perform certain mathematical computations – a task at which they excel. However, they do not understand the process in which they are engaged, nor the purpose of the exercise, nor even their own existence. In short, they lack perception and are unable to ask the questions that their human programmers can. Computers, at present, exhibit a low level of intelligence.

This state of affairs will not last indefinitely. As both our understanding of the human brain, and our ingenuity in manipulating microelectronic and micro-optical systems grow, the computer will acquire thinking powers increasingly like our own. Furthermore, aside from the unforeseen developments that are bound to occur, there is the future development of neural network computers, which, among other advances, could introduce the dimension of emotions into electronic intelligence – a matter to be discussed shortly.

Computers represent a wholly unique technology in so far as computer systems are capable of self-reproduction. It may not be cost-effective, but there is no theoretical reason why one could not create a fully automated factory, with self-correcting systems, which manufactures computers and robots without a human presence. As with any living organism, one would need to supply raw materials and energy to the factory and then eliminate waste products; one would also need to make provision for transporting away the computers and robots produced. However, these various processes could also be automated. In theory, one could even create and program such a factory complex to produce all the materials necessary for creating another factory, including the robots and automated machinery necessary for building the 'daughter' factory (for a detailed description of such a system, see Appendix C).

There may be good economic reasons for not creating such self-replicating robot factories now. There may even be good philosophical reasons in the future. On the other hand, it might become economically desirable if, for example, we wish to carry out mining operations on the Moon or beyond. Aside from speculating about the future, the fact remains that today it has already become virtually impossible to design the complex, miniaturized circuitry of advanced computer chips without the aid of computers. No steam engine ever designed another steam engine. In contrast, computers assist – and are vital in – the process of designing the next generation of computers. At what point will computerized expert systems take over the design of the next generation of computers from their human originators?

Finally, in the evolution of electronic intelligence, new characteristics can emerge and be inherited in a matter of milliseconds. This is why electronic intelligence will outpace human intelligence much sooner than we would like to admit. It took just 66 years from the time of the Wright brothers' first flight at Kitty Hawk to the time when men walked on the Moon, but in the history of technology there has never been a pace comparable to the advances in computers. For example, the power of memory chips sky-rocketed from 16 kbit to 16 Mbit between 1977 and 1992 – a thousand-fold increase in 15 years. Even if computers evolved no faster than aircraft, we should be on the electronic intelligence equivalent of the Moon by the year 2010.

The Emerging Global Brain

There is another, still more powerful form of electronic intelligence which will emerge during the 21st century. It is the collective intelligence of humanity wired together by various computer-based networks which will create an incredibly complex system analogous to the system of nerve fibres of the human brain.

To fully understand this process we need to remind ourselves of the concept of 'collective intelligence', a matter touched on in Chapter 4 and which has been more extensively explored in an earlier work (Stonier, 1992). Suffice it here to remind the reader of two examples: two people working together to solve a crossword puzzle are much more likely to finish the puzzle than one person working alone. Even more dramatic are the examples provided by social insects. The brain of an army ant has about a hundred thousandth of the number of neurons that the human brain has – it is an animal too stupid to survive on its own. However, 500 000 army ants, exquisitely integrated into a collective, become the terror of the forest.

The global networks which involve computers talking to each other, such as airline reservations, police vehicle identification, stock exchanges, weather forecasting, university networks and professional associations, will greatly upgrade the human collective intelligence. It will represent a process similar to earlier ones: when human speech first appeared, when writing and record-keeping became established, when a phonetic alphabet (Hebrew) was invented and then simplified by the Greeks, and when the printing press was introduced into Europe. That is, the global computer networks are facilitating the integration of global society into an ever more powerful collective intelligence.

In addition to improving the human collective intelligence, a second, entirely new, phenomenon will arise as we create a global communications network so powerful and so complex as to acquire properties not yet clearly understood. We will be creating, if unwittingly, an actual brain! The main difference will be that the network we will create will not be made up of nerve fibres, as is the human brain, but of telephone cables, optical fibres, radio waves and laser beams.

The majority of nodes of such a global brain will not be made up of neurons. Instead, they will consist of human individuals and groups coupled to their own personal computers. Each individual – as a logical outcome of the human need for social contacts – will be coupled to a number of friends and to thousands of other human/machine nodes for personal, business or social reasons.

This electronic global brain will be made up of an ever-increasing number of electronically integrated nodes, and, in due course, will also increase in specialized (but integrated) compartments. At the moment we already have such compartments: airline ticket reservation; global banking, postal and telephone systems; inter-library networks etc. In principle, this process does not differ from the evolution of primitive nervous systems into advanced mammalian brains: relatively few nerve cells, relatively poorly coordinated, evolving into an organ consisting of billions of cells so exquisitely coordinated that our understanding of how it works is only now, after many years of study, becoming comprehensible. With the evolution of the global brain we are dealing with a parallel process, but at a much higher level of complexity – in fact, with the upper end of known intelligence.

In addition to the basic anatomy, there are two major differences between the electronic global brain and its mammalian counterpart. First, the mammalian brain has a long evolutionary history behind it. It originated to serve the animal body, largely to coordinate the muscles and other bodily functions. The electronic brain will have a short evolutionary history and its main function is intellectual, not physical. Second, under normal circumstances, the nodes of the mammalian brain – its neurons – never duplicate themselves. In contrast, the nodes of the global brain – humans and computers – can increase almost indefinitely.

What will ***not*** be different when comparing the human brain with the electronic global brain is the magnitude of the complexity. Assume a global population by the middle of the 21st century of ten billion (10^{10}) people of which 10 per cent are likely to be included in the network. This would mean that the global brain would have a hundredth as many nerve cells as the human brain. However, most of the human nerve cells relate to the motor activities and physiological functions of the human body, which would not concern the global brain. Thus we should limit ourselves to the human cortex, which is estimated to contain only about ten billion cells (Shepherd, 1990).

Assuming that the number of connections to other nerve cells in the human cortex is of the order of 6000 – as it is in the visual cortex of the cat – this number is not out of line with the number of other human beings that individuals are likely to be in contact with during their lifetime. It is not the small number of family and friends, or even the much larger number of teachers and fellow students, but the hordes of shop assistants, waiters, taxi drivers, airline flight attendants, travel companions, hotel clerks and other 'strangers' with whom we interact, if only briefly, over the course of six or seven decades of modern life.

In the human brain there are at least two dozen neurotransmitters which cross the synapses that connect one nerve to the next. There are undoubtedly other substances which are part of the complex feedback mechanism that modulates and regulates the nervous system. But the same may be said for human social interactions. These include face-to-face contacts, with all their social subtleties; the use of the telephone, radio, television, faxes and email; and the numerous non-electronic communications systems, such as mail, trade and personal movements. The interactions between individuals in a global electronic society cannot be less complex than the interactions between neurons in the human brain.

The human brain has numerous 'organs' and 'tissues' designed for specialized functions, for example, the Purkinje cells, located in the cerebellum, which may be connected to as many as 200 000 other nerve cells. But consider a telephone exchange (mostly automated) which can handle hundreds of thousands of customers. As to the global brain's equivalent of the human brain's special organs and tissues – its pathways, columns, laminae, topographical maps etc. – think of all the libraries, archives and databases, all the universities and research establishments, the entire world's financial markets, the weather forecasting network, the air traffic control networks; we could fill many pages listing all the specialized institutions and organizations that are tied into, and are a part of, the global electronic network.

The nerve nets of the global brain, like their human counterpart, will be strengthened by use. For example, the communications channels between major cities are much greater than those between small towns. When there is an earthquake, say in California, friends and relatives telephoning from the east coast of the USA will not only comprise part of the market forces which encourage telephone companies to lay a sufficient numbers of cables, but many individuals will end up upgrading their telephone numbers, re-establishing contacts, and telephoning or writing at least once more following the disaster. Similarly, one consequence of the Cuban missile crisis of the 1960s was the creation of a 'hotline' between Washington and Moscow to facilitate negotiations between the leaders of the two countries.

An example of the global brain in action is illustrated by the following anecdote (Hayes, 1994; Taubes, 1994). Being able to send confidential messages from one computer to another is of major importance, not only for the military and other branches of government, but also for many commercial organizations, in particular those involved with financial transactions. Cryptographers are forever trying to create codes which cannot be broken.

Not surprisingly, there exists also a community of cryptographers as keen to break a new code as others were to create it. A single number, 129 digits long, known in the cryptographic community as *RSA-129*, became a major challenge. *RSA* refers to the authors who developed a code-making procedure: R. Rivest, from the Massachusetts Institute of Technology, A. Shamir, from the Weizmann Institute in Israel, and

L. Adleman, from the University of Southern California. The technique is based on multiplying two very large prime numbers (a prime number, e.g. 7, 11, 13, 17, is divisible only by itself and 1). It is relatively easy to find large prime numbers, and it is no problem to multiply two such numbers. On the other hand, to reverse this process, to factor the product, is extraordinarily difficult.

This system was developed in 1977 and RSA-129 was promptly put up to a challenge by Martin Gardner in his 'Mathematical Games' column in *Scientific American*. At the time, Rivest estimated that (given 1977 technology and mathematics) it would take 40 quadrillion years to come up with the two huge prime numbers which were multiplied to obtain RSA-129.

In 1993, a number of workers who were active members on an ***Internet*** mailing list devoted to cryptography and computer security decided to try to crack RSA-129 using newer technology ***and*** help from other ***Internet*** members. They managed to enlist 600 other individuals from 24 countries – all with access to their own and other computers (total 1600). These computers ranged from homey personal computers to a Cray supercomputer and several parallel supercomputers.

Instead of 40 quadrillion years, it took this form of distributed global collective intelligence only eight months to find the two huge prime numbers which, when multiplied together, gave the 129-digit number they were looking for.

A number of important lessons emerge from this experience: two are of particular relevance. First, it demonstrates the potential for using computer networks for obtaining both expertise and computer power from all over the world – a combination capable of solving a problem of incredible complexity. It means that, in the future, science and technology will be able to advance into areas previously thought impossible. Second, the 600 individuals which helped to factor RSA-129 selected themselves and, although coordinated through the Massachusetts Institute of Technology, worked on each part of the project with minimal central direction. In that sense they mimicked the activities of ants in an ant colony, or, more importantly, the neurons comprising the human brain.

The idea of a global brain is not new. In the middle of the 19th century, Nathaniel Hawthorne had one of his characters, Clifford, in *The House of the Seven Gables* express the following thought about the telegraph: '...by means of electricity, the world of matter has become a great nerve, vibrating thousands of miles... the round globe is a vast head, a brain, instinct with intelligence!'.

Early in this century, H.G. Wells called attention to what he termed 'the mind of the race'. In his day, terms like 'information technology' did not exist. It was long before computers appeared, and well before anyone was even thinking about information as an abstract entity. Yet he was thoroughly familiar with one form of information technology: ***literature***. Wells understood fully its significance: '...it is no doubt true that literature is a kind of overmind of the race...' (Wells, 1915, p. 167).

The Jesuit philosopher, biologist and anthropologist Teilhard de Chardin was perhaps the most articulate in forecasting the emergence of a single global intelligence. Influenced by the Russian geochemist V.I. Vernadsky (Krüger, 1981, p. 95), Teilhard considered the totality of humanity to represent '...nothing less than a 'sphere' – the noosphere (or thinking sphere) superimposed upon, and coextensive with... the biosphere'. To him, the ***noosphere***, a kind of collective global intelligence was: '...the final and supreme product in man of the forces of social ties'. From his biological perspective, he summed up the process leading to a global brain as follows: '...after six hundred million years, the biospheric effort towards cerebralization attains its objective' (Teilhard de Chardin, 1956, pp. 80–1). Teilhard also understood the basic difference in the anatomy and physiology of the human vs. the global brain: 'On the one hand we have a single brain, formed of nervous nuclei, and on the other a Brain of brains.... Whereas in the case of the individual brain thought emerges from a system of non-thinking nervous fibers, in the case of the collective brain each separate unit is in itself an autonomous center of reflection' (Teilhard de Chardin, 1959/1964, p. 173).

Neural Network Computers

Neural network computers (NN-computers) represent the natural next step in the evolution of electronic intelligence. It was realized, decades ago, that the human brain processes information in a manner quite different from the way that a conventional digital computer does. This provided the motivation for creating, electronically, something which mimicked more closely the ways the human brain worked. In NN-computers, as in the neural networks of the human brain, connections become strengthened by use. Also, like the flip-flop circuits of the thermostat system discussed earlier, firing one circuit inhibits neighbouring circuits and, in turn, weakens their connections.

Neural network computing is still largely in its infancy. By and large NN-computers are still, primarily, devices found in universities and other research centres, where they got their start. The fundamental theory of neural computing was first explored in the early 1940s by W.S.

McCulloch and W.A. Pitts at Chicago. In the early 1950s, B.G. Farley and W.A. Clark created models for adoptive stimulus–response relations in random networks. These concepts were developed further by a number of workers – Rosenblatt, Widrow and Hoff, Caianiello, and Steinbuch.

Some workers, e.g. Obermeier and Barron (1989), consider Bernard Widrow, working at Stanford University in the 1950s and 1960s, to be the founder of the field of neural networks. One of Widrow's early neural network applications was a weather forecasting system. By feeding the network many samples of yesterday's pressure and today's weather, it came up with tomorrow's forecast when fed today's pressure. The accuracy of the local meteorologist was about 65 per cent – that of Widrow's network about 83 per cent.

Kohonen (1988) estimates that, during the quarter of a century between the early 1960s and the late 1980s, a few thousand papers have been published on general neural modelling. He defines 'artificial neural networks' as: 'massively parallel interconnected networks of simple (usually adaptive) elements and their hierarchical organizations which are intended to interact with the objects of the real world in the same way as biological nervous systems do'.

Even if, as yet, we cannot remotely hope to achieve computers whose complexity approaches the human brain – with 10^{11} transputers involving, perhaps 10^{15} connections – the potential of neural network computers is enormous. Although it will be some time before computers achieve the complexity of the human brain, when processing information electronic circuits have one advantage over their biological counterpart: ***speed***. Events in a silicon chip take place in the nanosecond (10^{-9} s) range while neural events take place in the millisecond (10^{-3} s) range – electronics can process information a million times faster than can nerves. (A comparison between conventional digital computers and the human brain is provided in Appendix D.)

In reviewing the advantages of neural networks, Haykin (1994) lists nine properties and capabilities. It is not the purpose of the present work to relate these in any further detail other than to comment that these properties presage a revolution in electronic intelligence, a revolution which may be analogous to the improvement in the vertebrate nervous system over the invertebrate system which preceded it. Sometime early in the 21st century electronic intelligence will undergo a phase shift as a result of the further development of NN-computers. At the moment, to quote Haykin (1994, p. 7): 'The artificial neurons we use to build our neural networks are truly primitive in comparison to those found in the brain'. However, Haykin continues: 'What is really satisfying... is the remarkable progress that we have made on so many fronts during the past ten years... and... it is for certain that in another 10 years our understanding of artificial neural networks will be much more sophisticated than it will be today'.

Even more positive is Robert L. Harvey of MIT's Lincoln Laboratory, who considers that: '...human consciousness may be the last surviving mystery because people do not know how to think about it'. Harvey concludes that using NN-computers to mimic consciousness and higher level mental processing '...may be the next major milestone' (Harvey, 1994, pp. 175, 178).

Summary and Conclusion

It is not the purpose of the present work to delve deeply into the workings and further potential of neural network computers. The point is that NN-computers represent a facet of electronic intelligence which, when further improved and coupled to classical digital computers, will result in computers that are smarter than people. Thus, we have witnessed a process during our lifetime which began (quite innocently) with humanly created proto-intelligent systems and evolved into truly intelligent systems with the promise that these systems will, in turn, evolve into super-intelligent ones.

The emergence of electronic intelligence is important not only in terms of the profound social changes it will bring about, but also from a theoretical standpoint: its uniqueness in the cosmic evolution in our part of the universe. As Forsyth and Naylor (1985) have pointed out: 'Humanity has opened two Pandora's boxes at the same time, one labelled genetic engineering, the other labelled knowledge engineering. What we have let out is not entirely clear, but it is reasonable to hazard a guess that it contains the seed of our successors'.

Finally, as a part of considering the progress of electronic intelligence, we may ask: at what point will human commands become ***meaningful*** to a computer? It is a question which could be considered as a major theme of the rest of the book.

· 8 ·

The Physical Origin of Meaning

Introduction

Imagine an affluent home in which the lady of the house has engaged a cleaning service to clean the house. The cleaning service has sent a supervisor and several labourers to carry out this task. The labourers are immigrants from a country whose cultural patterns are substantially different from the Western culture dominating the home.

While cleaning the living room, one of the labourers discovers a piece of a jigsaw puzzle under a chair. Since jigsaw puzzles are not a part of his cultural heritage, he has no idea what it is he is looking at. He is about to throw it away when he decides that this strange and apparently useless object probably is of some use to somebody. He therefore takes it to the supervisor.

The supervisor instantly recognizes it to be a piece of a jigsaw puzzle and takes it to the mistress of the house. The mistress explains that her son Johnny had been working on such a puzzle in the living room the evening before but had moved it to his room to get it out of the way of the cleaners. She therefore takes it upstairs to John's room, where she knows he is still working on it.

Johnny has practically finished putting the picture puzzle together when his mother appears with the piece lost on the living room floor. Without looking at the puzzle, Johnny recognizes the piece instantly and exclaims: 'Oh! That's where that piece of sky went'.

Now, among all the physical objects one can think of, a piece of a jigsaw puzzle is about as useless an object as one can imagine. It is not edible, nor will it allow you to fry an egg, dig a trench, clean the house, invest in the stock-market, nor will it help much with any of the ordinary routine problems which govern our daily lives. On the other hand, if a picture is worth a thousand words, then ***a picture must contain a message.***

And if a picture contains a message, then a ***piece*** of the picture must be a piece of information.

Note how the same piece of information, the piece of jigsaw puzzle, was interpreted differently by the four people involved. For the labourer the piece had no meaning other than that it was some sort of artefact – that it was not a piece of dust or trash and that it was not an object to be discarded. For the supervisor it was more than a mere artefact: it was a part of a category of things called jigsaw puzzles. For the mother, it was a part of a ***specific*** jigsaw puzzle – one she had seen in the living room; and for Johnny, it was not just any piece of jigsaw puzzle: it was a specific part of his puzzle – even though he had never seen it before, it was a piece whose shape and colour allowed him to recognize it instantly. As the puzzle piece moved from labourer to Johnny, the same piece of information became increasingly more meaningful.

The Spectrum of Meaning

The above thought experiment demonstrates that the same piece of information may increase in meaningfulness depending on who looks at it. This opens the door to the possibility that 'meaningfulness' might be quantified. Such a possibility may not be realizable as yet. However, it should allow us, at a minimum, to define the problem more precisely. We shall return to this matter later, but before we do we need to explore the idea, already suggested by the above, that just as there exists a spectrum of intelligence, so does there exist a ***spectrum of meaning***. At the high end, meaning leads to understanding and further analysis, while at the low end, meaning involves something as simple as the ability to pick out an item from among an array of items.

Let us begin with another thought experiment, but this time with a system much simpler than one involving human thought and action. Consider an ant foraging for food. As it moves around, it finds a grain of sand which it ignores; the ant then runs across a piece of wood which it ignores as well. It moves on and encounters a piece of dirt. Again it ignores it. But then the ant comes across a grain of sugar which it explores, then picks up and drags back to the nest. Question: does the grain of sugar have 'meaning' for the ant?

Clearly, the ant is able to ***discriminate*** between various items it encounters during its search. Such acts of discrimination imply that the grain of sugar has ***meaning*** for the ant. However, this does not mean that the ant ***understood*** either the ***significance*** of the sugar, or that of its own actions. An ant brain contains only of the order of 10^5 neurons. Presumably, the ant was operating on simple sets of algorithms programmed into its brain. For example, the algorithm may be simply this: if it tastes sweet, eat if you are hungry, take it back to the nest if you are not. Algorithms of this sort can regulate the behaviour patterns of simple robots.

Consider another scenario: the ant meets another ant. By means of pheromones the second ant sends the first a message. On the basis of that message, the first ant alters its behaviour. Here we have a traditional semiotic situation of ***sender–message–receiver***. Since the receiver, the first ant, was able to interpret the message and act on it accordingly, the message must have had 'meaning' for the ant.

Did the first ant '*understand*' the message? Probably not. Probably the ant acted on another set of algorithms programmed into its brain. As alluded to in an earlier chapter, individual weaver ants perform probably no more than 50 distinct behavioural acts, of which the majority involve acts of communications (Hölldobler and Wilson, 1977). It would not require a very large number of algorithms to cover the entire behavioural repertoire of an ant – certainly well within the capacity of a brain consisting of 10^5 neurons.

If a grain of sugar has meaning for an ant, to be consistent, does that not mean that for an ***amoeba***, a morsel of food has some sort of meaning? As with the ant, the amoeba is able to discriminate between food and non-food. Likewise, for a ***human leucocyte***, which, as it circulates in the bloodstream, is able to discriminate between benign native and invasive foreign entities. Do these 'white cells', therefore, not derive ***meaning*** from entities encountered in the bloodstream? They respond as though they have received ***and interpreted*** messages – for they act accordingly: attack or not attack.

Would the same not apply to ***antibodies***? Although these immunoproteins lack all cellular organization, they are able to ignore a vast number of substances circulating in the bloodstream but attach themselves to those considered to be antigens: foreign substances which have no business circulating in the bloodstream or lodging in the tissue. This ability to discriminate (and antibodies are usually highly specific) would suggest that, for an antibody, an antigen has meaning.

This is not a trivial thought. An increasing number of immunologists are drawing a parallel between the immune system and language (Tauber, 1994, pp. 168–72). What is not resolved at this point is whether the flow of information leading to recognition represents a system of communication – a semiotic interaction – or simply a sophisticated stimulus–response pattern. This question is misleading. Within the framework of the present conceptualization, i.e. based on the notion that information-processing systems have evolved and now present us with a spectrum of interrelated phenomena, stimulus–response patterns must, perforce, also involve a semiotic interaction (although it may be of a much less sophisticated kind). The question therefore should not be: 'Do the various interactions associated with the immune system and the immune response involve semiotic interactions – interactions in which the recipient of information ascribes 'meaning' to the incoming information?', but rather: '***How much meaning*** to the incoming information can such a system confer given the limits of the ***internal information environment*** provided by the antibody, or even, the cellular phagocyte?'.

The antibody–antigen interaction suggests an even broader question: does a metabolite which constitutes a specific substrate for an enzyme have ***'meaning' for that enzyme***? The enzyme may be confronted with thousands of different kinds of molecule, yet it may distinguish only one specific kind of a molecule as a substrate molecule. To be consistent with the above, the answer must be: *yes* – a substrate molecule must have meaning for an enzyme.

This conclusion becomes less controversial if we recognize that, when the enzyme engages in an act of selection, the selection is based on a form of information processing. The choice on the part of the enzyme relates to the structural and electronic properties of both the enzyme and the substrate. It is these properties that allow for a close fit of the two molecules. Basically, therefore, identification is based on ***tactile*** processes. At the molecular level, these are based on a variety of physical forces: electrostatic forces, van der Waals forces, hydrogen bonding etc. Undoubtedly, the 'tactile' response also involves the establishment of electronic resonances. For the enzyme, the pre-existing organization of the protein allows these tactile interactions to occur. This organization constitutes the enzyme's ***internal information environment***. It is this information environment into which the substrate nestles and which, by our operational definition, becomes a 'meaningful' stimulus for the enzyme. The substrate molecule constitutes the message, the enzyme provides the context. Keep in mind that usually the enzyme molecule is hundreds if not thousands of times larger than the metabolic substrate.

To strengthen the supposition that we are dealing with semiotic interactions, the enzyme molecule itself may be subjected to further regulation by other ***chemical messages*** emanating from within the cell or from outside. Such messages are able to change the structural organization of the enzyme, thereby inhibiting its actions or enhancing it. Thus the regulatory message may signal the enzyme 'do not attack' (or attack with less efficiency). Alternatively, the regulatory message may be an activator and enhance the 'attack' message. What is important in all this is that the regulatory message is a molecule which attaches itself to the enzyme and, by changing its organizational structure, changes its internal information environment. If the enzyme has become totally inhibited, the internal information environment would have become so altered that the enzyme would no longer interact with its substrate. The substrate would have become meaningless to the enzyme. We will explore the concept of the internal information environment in greater detail in the next chapter.

Finally, a ***crystal*** growing in a saturated solution of the crystalline material discriminates between crystal molecules and solvent molecules; or in an impure solution, discriminates between crystal-forming materials and impurities. This brings us down to the lower end of the spectrum: we are able to ascribe meaning to processes occurring in inorganic systems.

The above line of argument becomes more palatable if we recall that all of the above interactions involve existing information-processing systems. Notice how in this analysis we have divorced the concept of 'meaning' from the concept of 'understanding'. We shall put off exploring this distinction until the next chapter.

Meaning, therefore, represents a spectrum of phenomena. At one extreme is the purely mechanical process exhibited by a growing crystal selecting the appropriate atoms or molecules out of a mixture of substances which make up its environment. Slightly higher up the scale of complex behaviour is an enzyme molecule which discriminates from among hundreds (if not thousands) of molecules, to enfold a specific substrate molecule. Like a radio receiver, the enzyme itself may be 'tuned' by the internal information environment of the cell. That is, the cell, in order to achieve the coordination of its metabolism, activates or inhibits an enzyme by complexing it with other molecules which function as chemical messengers. These secondary molecules also have a meaning for the enzyme: the enzyme provides a specific site of attachment – a site that is highly organized and hence rich in information. This information represents the internal information environment of the enzyme, which allows the message to be decoded appropriately. (This is as true for competitive inhibition as it is for allosteric inhibition.) If the message results in a sufficiently strong inhibition, then the substrate molecules will no longer have any 'meaning' for the enzyme: the enzyme will have become 'tuned out'.

At the other extreme of complexity one might contemplate a philosopher pondering the meaning of meaning. Here, the original message–context complex appears as a new piece of information to the brain, treating it as a new (internal) message to be placed into a second context. This second-order meaning complex is further information processed into a third order, and so forth. The philosopher contemplating the term 'meaning' will have reached an *n*th-order complex, a level which, given our present state of knowledge, we are unable to define.

The important point is that ascribing meaning to a message involves a ***spectrum*** of phenomena. At the lower end of this spectrum, the process of ascribing meaning to an incoming message or other form of information does *not* require intelligence. At the upper end, the internal information environment presented by the brain of the philosopher is so rich that it allows the incoming information to be compared to many different aspects of pre-existing and organized information, which, in turn, allows meaning to be extracted at many different levels, resulting in the formation of new knowledge structures.

Defining Meaning

The dictionary definitions of 'meaning' in general imply that only human beings are able to ascribe a ***meaning*** to anything (symbols, words, sentences, messages, speeches, books, actions, thoughts, events, situations etc.). Such definitions presuppose that only human beings are capable of ascertaining the meaning of things: no references are made to ***non-human*** mental processes, much less to those processes which involve no mental activity at all.

To develop a theory of meaning we must first rid ourselves of the anthropocentric idea that only humans are able to ascribe meaning to incoming information. Second, we must differentiate clearly between 'meaning', 'ascertaining meaning' and 'understanding'.

Meaning is a ***relationship*** between information which exists in the environment of a potential recipient of such information, and the ***internal information environment*** which already exists within the recipient. When these two separate and distinct sources of information become complexed, a meaningful ***state*** has been achieved. The idea of an internal information environment is a most important concept and will be explored in greater detail in Chapter 10. Suffice it here to say that it is the information stored in a recipient which links up with the incoming information (such as the human brain, or the DNA of a cell, or the active site of an enzyme). Once the external information has become integrated into the internal information environment, the recipient system, if it is sophisticated enough, is able to engage the complex in further information processing.

To make the above more comprehensible, we might return to our foraging ant as a metaphor. The ant, as it crawls along, discriminates among the various objects it encounters in its environment. It ignores the sand, wood and dirt – it establishes no relationship with these objects. The grain of sugar, on the other hand, elicits a more positive relationship. In our thought experiment, let it be a hungry ant. It not only examines the grain with interest, it begins to devour it.

Consider the inside of the ant as representing its internal information environment. Inside this internal environment the sugar grain begins to be processed. First it is dissolved – the individual molecules are separated. Then the molecules are processed to yield either energy or building blocks for the ant's tissues. Within a matter of hours there is no trace left of the grain of sugar. It has been digested, distributed and entered into the metabolic pathways of various cells. It has become an integral part of the ant.

Anyone who has worked or played with young children will recognize this metaphor. Young children have this wonderful gift of sweeping up all sorts of information that comes their way. They are like a powerful vacuum cleaner – everything becomes funnelled in. This new information gets 'digested' by the child. First it becomes sorted into categories, then, in due course, it becomes incorporated into the child's knowledge structures. This is how children create an increasingly complex model of the world, a model which allows them to operate in their environment with increasing sophistication.

There can be no meaning without a context. ***Meaning*** may be defined as the capacity of an intelligent or proto-intelligent system to discriminate between various aspects of environmental information and act upon an appropriate stimulus in a functional manner. In the case of proto-intelligent systems, such as crystals, enzymes and antibodies, functionality consists of carrying out the appropriate chemical or crystallographic processes for which the system was designed. In intelligent systems, functionality consists of behaviour which is intelligent (i.e. behaviour which enhances survival or reproduction).

The above conceptualization of 'meaning' overlaps with the definition of 'information' used by library information scientists. For example, Brian and Alina Vickery (1992, p. 30) present the following scheme: 'A source, S, emits a signal or message, M, from which information, I, is assimilated by a receiver, whose state is consequently changed from (K1) to (K2)'. The Vickerys imply throughout their book that only the information (I) which is assimilated by the receiver and has an impact on that receiver is true information. In the present work there is a subtle but important distinction made: what the Vickerys call 'information', to be precise, should be called 'meaningful information'.

Let us take the example of a statement such as:

'*Nach mir die Sintflut*' he recited, looking aggressively at his fellow inmate.

For the average English-speaking person who has no knowledge of German, only the second part of the sentence contains meaningful information. The first part, although it is not totally meaningless, does not conjure up images, as the second part does. The entire statement contains information! However, only the second part resonates within the internal information environment of the English-speaking person.

This brings up another aspect of our analysis, an aspect which, so far, has been ignored. In the above statement, what was the intended meaning of the author when constructing such a message? How do messages achieve their meaning?

Here we must differentiate between two kinds of message. The first are messages or signals sent out for reasons other than to communicate. These are 'unintentional' messages which, in evolutionary terms, preceded the second kind. They will be the subject of our discussion later on. The second kind are 'intentional' messages, sent out with a view to communicate with some recipient. These messages are characteristic of social organisms – keeping in mind that we are dealing with a spectrum of phenomena. The 'mating' behaviour of two bacterial cells lining up to engage in an exchange of genetic material is not on a par with an ant secreting a pheromone to induce a specific activity in another nest mate, which is not on a par with the social communication between two monkeys, which is not on a par with an author attempting to communicate with a reader.

Intentional messages sent by a sender always take cognizance of the receiver's internal information environment. This may not require a conscious act, as in the case of the bacterium or the ant. It may simply have evolved as part of a communicating (information processing) system. It is, however, always true that the sender tailors its message so as to be meaningful to the receiver. This is as true for a bacterium releasing chemicals to line up another bacterium with which it will exchange genetic material as it is for the ant secreting a pheromone, or for a primate signalling a conspecific, or for an author trying to reach a reader. An author aiming for an English-speaking audience, for example, would not write in German and would not, generally, even interject a German expression without putting the translation into a footnote.[1]

In humans, intentional messages designed to be meaningful to a recipient rely on the internal information environment of the sender to give meaning to the message. The larger and more sophisticated the internal information environment is, the greater is the possibility of

1 '*Nach mir die Sintflut*' translates into 'After me the deluge'.

tailoring the message so as to resonate with the internal information environment of the recipient. This is the struggle that authors engage in to make their writings more 'meaningful' to readers. Our analysis of 'meaning' therefore is as applicable to the sender as it is to the recipient. Meaning is a function of context. In humans, for both sender and receiver the context is determined by their respective internal information environment.

As to message, intentional messages are never 'naked' messages. They are messages always sent and transmitted in some sort of contextual bed. This contextual bed frequently becomes part of the 'meaningful' information reaching the recipient. That is, the context surrounding the message – where it appears, when it appears, under what circumstances it appears, who the sender is, who the messenger is etc. – becomes 'meaningful' information, i.e. these factors are incorporated into the internal information environment of the recipient, where they are further processed.

Thus, whether one wishes to ascertain the meaning of a message in terms of the intent of the sender, or whether one wishes to ascertain the meaning of a message in terms of its information value to the recipient, each can be accomplished only within the framework of the internal information environment provided by either.

Whereas 'meaning' is a ***state***, 'ascertaining meaning' is a ***process***. Specifically, it is the process of attempting to place information into a ***context***. It involves a type of information processing by which the external information coming in is matched up with the internal information environment of the recipient. If the ***message*** is able to resonate with some part of the internal information environment (the context) it will have meaning for the system; if not, it will be ignored.

Which Came First: the Message or the Meaning?

The idea that intelligence is not required for achieving meaning and that meaning involves a spectrum of phenomena allows us to move back in evolutionary time to understand the origin of codes.

In Chapter 6 we addressed ourselves to the origin of bio-intelligence. One of the theories we discussed was the 'clay hypothesis'. If this theory proves to be true, even if only partially, one of the important consequences of such a theory is the idea of the 'naked gene'. For a clay gene, the message existed first and the meaning came later.

Even if the clay hypothesis proves to be wrong, the principle of the naked gene is probably correct because it must have been equally true in the pre-RNA world. The original RNA polymers initially must have represented a code without meaning. In each case, there appeared a self-replicating template which acquired meaning only after it interacted with its environment in an organized manner. Ultimately, such a micro-environment evolved into an ***internal information environment***, but only after the whole system became encapsulated within some form of membrane system. Only then could a genetic code begin to have meaning for the biosystem of which it had become a part.

Once biotic systems had become well established and ***life*** as we know it was interacting with the environment, a new set of codes emerged. Such a set involved information coming in from the environment which was associated with some other environmental trait of importance to the survival or reproducibility of the recipient organism. Two examples will suffice.

Many primitive animals, including protozoa, have 'eye spots' which guide them towards light. Although such animals do not photosynthesize, and hence do not use the light directly for purposes of growth and reproduction, the light represents a cue as to where food is likely to be found. In other words, for a ***Paramecium*** light constitutes a message which when decoded means: 'food over here'.

For a seed germinating in dark soil, gravity represents an important source of information which orients the axis of growth. However, for a plant, gravity is neither a useful source of energy nor a source of nutrients. Yet it is by using gravity that the plant sends its shoots up to the light and its roots down into the soil. Gravity therefore has meaning for the plant and represents a message which, when decoded, allows it to find energy above and nutrients below.

Here, again, note how the potential messages (***light*** in the first instance, ***gravity*** in the second) existed long before they became recognized by biosystems as such. We must conclude, therefore, that at the very beginning, messages existed first and that meaning, and means of decoding these messages, came second. Furthermore, what ultimately became a code had its origin in naturally occurring phenomena which involved neither intentions on the part of the sender (the Sun's light, the Earth's gravity), nor meaning on the part of the recipient. So, like the concept of the naked gene, the information was there long before it became meaningful.

Also note how, in the two examples cited, the messages did not require a brain for them to be decoded. In fact, they did not even require a nervous system. Even in animals that do possess a brain, there are innumerable examples of a precursor message being well established before it achieved a meaning. To name a few: the poison glands of ants serve as a method of defence and for subduing prey. However, in some species, a few constituents of the poison glands have evolved to serve as recruitment or alarm substances (Hölldobler and Wilson, 1990, p. 239). Another example, among weaver ants, is that following an encounter with alien ants an excited worker alerts nest-mates to the danger by greeting them with a rapid back-and-forth jerking movement. This movement is thought to have evolved as a ritualized version of attack behaviour (Hölldobler and Wilson, 1990, p. 252).

This principle, of body movements initially designed for non-communicative purposes, such as attack, defence, food-gathering or grooming, evolving into a secondary function of communication, was well established during the 19th century when it occupied minds as profound as those of Charles Darwin and Herbert Spencer (Andrew, 1965). It is probably true to say that virtually all body language used for communication derived from some primary bodily function which had nothing to do with communication. This includes facial expressions, which were derived from reflex reactions such as biting, sucking, flattening of the ears etc.

To sum up: from naked genes in pre-biotic systems to the extensive body language exhibited by higher primates, the seed of the message came first. Only later did it come to acquire meaning. As we shall discuss in the next chapter, the origin of human speech itself had its antecedents, without which it could never have emerged.

· 9 ·

The Origins of Human Speech

Introduction

Human speech as we know it emerged only after a complex evolutionary process. Philip Lieberman (1988), Professor of Cognitive and Linguistic Sciences at Brown University, has reviewed the incremental mutual stimulation and integration of brain development and vocal anatomy which occurred during the past several hundred thousand years, and which must have preceded the full flowering of human speech.

Among primates, only humans possess the anatomical structure required for the full range of human speech sounds. This ability reflects the location of the voice box – the larynx – which is located deep in the throat of adult humans. In contrast, inarticulate animals, including apes and human infants, have their larynx high in the throat.

Unlike human adults, every other mammal can breathe and drink at the same time. This useful facility is brought about by a larynx high in the throat which connects the nasal passage directly to the windpipe, allowing food and water to bypass it on the way from the mouth to the stomach. This blessing, of being able to drink and breathe at the same time, is extended to human infants for the first few months of their lives, when they have not, as yet, much to say for themselves.

In contrast, human adults, whose larynxes have retreated down into the throat, choke to death by the hundreds each year. Nature is never as careless as all that: there would be no selection in favour of some deleterious trait unless that trait proved to provide a strong selective advantage in some other way. In the case of the descending larynx, it is clear that the organ which, in mammals, was primarily designed to facilitate drinking, and only secondarily to act as a voice box, became modified in humans so that the relative importance of these two functions became reversed.

Obviously, for this to occur there must have been strong selective pressures favouring the ability to modulate sound.

It would seem highly improbable that such selective pressures would have been very strong if there hadn't already existed a well developed system of communication and signalling. Such a system would have consisted of two major and distinct forms of social technology. The first, and the one usually considered the more important in the development of human speech, is based on vocalization. The second is based on visual signalling. Let us examine these two.

Vocal Signalling Among Vervet Monkeys

Perhaps the best examples of communication by vocalization are the comprehensive studies on vervet monkeys (*Cercopithecus aethiops*) by Thomas Struhsaker and colleagues (Struhsaker, 1967). Struhsaker describes and catalogues the calls, cries, squeals, barks, grunts, purrs, chutters, teeth chattering, lip smacking and other sounds which comprise the vocal repertoire of these monkeys. It is apparent that these sounds are emitted in the context of different situations and constitute meaningful messages to conspecifics. The meaning of each sound is, of course, inferred by the behavioural response of the recipient(s) to whom the message is addressed. The call may be a natural one produced by an individual monkey, or it may be a tape recording played back by a scientist. The recipient might be a single individual such as a mother responding to her infant's 'lost squeal' or a young male giving way in response to his superior's bark. On the other hand, it may involve the whole troop responding to an 'Aarr' call, indicating the approach of a group of neighbouring vervets, thereby setting the stage for a possible territorial confrontation.

Of particular interest are the alarm calls alerting the troop to a predator. Such predators include eagles, leopards, serval cats, lions, cheetahs, hyenas, jackals and hunting dogs. Humans and snakes also may pose a danger. By using tape recordings of various calls, it was possible to show that the alarm calls for these various forms of danger were different and elicited entirely different responses.

For example, the 'snake chutter' caused the monkeys to respond by forming a cluster about five feet from the snake, staring at it while continuing to emit the call. As the snake crawled through the thicket, the vervets would move along beside it, continuing the chutter. The snake chutter is a soft call designed to warn others nearby without attracting the attention of other, more lethal, predators.

The 'Uh!' call indicates the presence of minor mammalian predators, such as hyenas, jackals, cheetahs, hunting dogs and, at times, Masai tribesmen. It is a low-intensity call which causes the others to look towards the source of the alarm call, then either ignore it or move away.

Quite different is the response to the 'Rraup' call announcing the sighting of an eagle. The call is of short duration and is almost never repeated. Immediately, the entire troop retreats into thickets, either by dropping down from the trees or by running for cover on the ground.

If the bird of prey circles or attacks, the 'Rraup' call may be followed by threat-alarm-barks and chirps. Chirps and threat-alarm-barks would also be issued when serious mammalian predators appeared on the scene: leopards, serval cats and, to a lesser extent, lions and, in some parts of Africa, humans. These calls would cause the monkeys to scamper up into the trees.

The threat-alarm-bark appears to have two functions: first, to act as a general alarm for the rest of the troop and second, to threaten the predator or invading vervet neighbours. It was Struhsaker's impression that the closer the threat, the louder the call and the more rapidly repeated were the phrases making up the call. In this manner it would be possible to convey to the others how imminent the threat was.

These examples of vervet vocal communication (which also provide an excellent example of ***zoosemiotics***) make most plausible the idea that human speech did, indeed, simply evolve from an extension of the more primitive, primate vocalization. However, to the present author it seems that much more would be required. For example, it is not clear how animal howling would evolve into ***consonants***. It seems much more likely that, in order for the late hominid brain to evolve the faculty of speech, a number of other pre-existing communication skills were combined to create truly human speech

Visual Signalling

One such communications system, well established among all higher primates, is ***body language***. Body language includes facial expressions such as raised eyebrows, frowns, smiles and narrowing of the eyes (Andrew, 1965). The subject was put on a firm foundation when Charles Darwin published *The Expression of the Emotions in Man and Animals* in 1872. Facial expressions evolved to convey moods and to signal intentions.

Such a signalling system could be easily extended to convey other types of information as well. Gesturing 'come here' or 'follow me' or 'stay there' – commands easily taught a dog – would be one form of extending body language. Another would be ***pointing***. Being able to point silently to a source of danger not readily visible (a leopard or a snake), or to a more distant source of food or water, would have survival value for a social animal such as a higher primate. If the resource in question were out of sight, let us say on the other side of a hill, the pointing gesture might be modified to include an undulating motion using the forearm and hand to mime going up and then down that hill.

If such a gestural language did develop, the young would learn it by imitating their elders – that is, by ***miming***. Such miming would not always be confined to juveniles. The present author observed a classic case of miming at the San Diego Wild Animal Park when a bonobo named Loretta tried to gain the attention of one of the keepers. After the usual attempts – presenting, then presenting while peeking between her legs – she faced him and began nodding her head. Loretta had learned this behaviour by watching humans talking to each other. In Western cultures, when two people engage in conversation, the silent partner will usually nod to encourage the speaker.

Miming is still, today, a universal human language. An English person may not understand a word of French, yet will have no problems understanding the antics of a world-class mime such as Marcel Marceau. Likewise, the Australian aborigines are noted for their ability to mimic the animals they encounter during their hunt.

The idea that human communication was, at its inception, primarily gestural rather than vocal, has a long and honourable history dating back to the 18th century. In reviewing the matter, anthropologist Gordon Hewes (1973) concluded that: 'a pre-existing gestural language system would have provided an easier pathway to vocal language than a direct outgrowth of the "emotional" use of vocalization characteristic of non-human primates'.

The question which arises is this: how might a gestural language give rise to a vocal one? The answer assumes that, superimposed on the existing primate (vocal) call system, there was a second system. This second system created 'words' by using the mouth parts to imitate the movement of the miming hands and arms. This idea has been around for many years. Alexander Jóhannesson, Professor of Comparative Philology at the University of Reykjavik, studied words from ancient Icelandic and from Hebrew, the former an Indo-European language, the latter a Semitic one. Citing his own studies and the work of the English linguist Sir Richard Paget, Jóhannesson considers 'fully proved' (Jóhannesson, 1949,

p. 82) the theory that many words originated in the primordial human language from the mouth parts mimicking what the hands were doing during miming. The reader is invited to experiment by miming with the hands, words like 'round' and 'hole', 'flat' and 'plain', and noting the corresponding movements of the mouth parts.

This experiment should not be construed to mean that these precise words are the same as the original human language (proto-World) which arose in East Africa. After all, at least a hundred thousand years of linguistic evolution separate contemporary English from proto-World. (Consider how, in a matter of mere centuries, the now almost incomprehensible English of Chaucer has evolved into contemporary English.) However, the experiment does demonstrate that the origins of human speech probably derived not only from animal calls but from gestural language as well.

Jóhannesson's theory is strengthened by the observation that language capability resides in that brain hemisphere which controls the side of the body preferred for posturally demanding tasks (such as balancing on a shaking limb while reaching for food). Some linguists (MacNeilage et al., 1993) have hypothesized that: 'the left hemisphere of the brain, already primed for controlling posture, gradually came to control the communication skills, which were posture-dependent'. The muscles of the mouth, face and neck would be included in the control of posture.

These ideas relate to those of Charles Darwin, who pointed out that difficult hand and finger movements were often accompanied by thrusting out or twisting of the tongue, or by pursing the lips, as when threading a needle.

Group Vocalization

Group vocalization must be considered as another language precursor. Social carnivores, such as wolves and jackals, engage in group howling. Judging by the vigour of their tail-wagging, these appear to be extremely pleasurable activities. All human cultures engage in some sort of drumming and group vocalization, usually accompanied by dancing. This too seems highly pleasurable for the participants. Group shouting and the use of drums to frighten off predators or stampede game may have been a quite ancient custom among hominids. It is also quite possible that group vocalization – as with howler monkeys – may have played a role in territorial confrontations.

The use of drums and group vocalization may have evolved into formal rites – strongly selected for because of their biological value in integrating group behaviour. The same could be said for rites which involved group dancing.

All languages have a tune to them. These so-called ***pitch contours*** are learned by babies before they have uttered their first word (Moskowitz, 1978). By the late babbling period, children are able to control the pitch contours of the language they hear around them. At that stage children may sound as if they are uttering reasonably long sentences, sound sequences which fool their adult listeners into believing that the baby is actually talking – even if they could not quite catch the words. In fact, there were no words – only a tune to the babble.

In due course, the babble disappears as it begins to be replaced by actual words. The tune, however, is retained. The very early acquisition of a language's pitch contours may explain why adults frequently have such difficulty in learning the intonation of a second language. In other words, pitch contours are one of the aspects of learning a language which involve the ***imprinting*** phenomenon discussed earlier.

Finally, the fact that an entire tone system is mastered while the baby is still babbling leads to the interesting thought that, during the emergence of human speech, the earliest morphemes may have been nonsense syllables which were shouted as cadences or sung as a part of ancestral dancing rites. We may still hear such cadences when we observe primitive rites such as the cheerleaders at an American football game: 'Rah, Rah – Siz, Boom Bah!'.[1]

Laughter

Another form of group vocalization is ***laughter***. We all know that laughter can be infectious, a fact used by radio and television shows that add

---

1 The use of nonsense syllables in song is still common around the world. Here are a few examples in English:

1. Tra-la-la: in children's literature to indicate that a character is singing.
2. Taradiddle-taradiddle-tol-lol-lay: Gilbert and Sullivan opera.
3. Turah-lurah-lay: part of a well-known song performed by Bing Crosby.
4. Hey-diddle-diddle: nursery rhyme.
5. Hi-ho-the-derry-oh: folk song.
6. Tarrara-boom-dee-yea: march.
7. Mersey-doats-and-dosy-doats: pop song of the 1940s.
8. Hut-sut-ralston-on-the-rillerah: ditto.
9. Fallalala-lah lah-lah lah lah: Christmas carol.

'canned' laughter to stimulate audiences to laugh. Sometimes, uncontrollable laughter can reach epidemic proportions, such as the 1962 outbreak of contagious laughter in Tanzania, which laid whole villages low (Provine, 1996). Robert Provine, Professor of Psychology at the University of Maryland, Baltimore County, suggests that human beings have an 'auditory feature detector', a set of neural circuits that responds specifically to human laughter. These circuits trigger other neural circuits leading to the recipient's unconscious response to join in laughter. If Provine is correct, much laughter involves a reflex reaction.

Interestingly, chimpanzees and other great apes perform laugh-like vocalization when tickled or during play. However, the respiratory pattern of chimpanzees during laughter is very different. Human laughter, which produces a series of sounds such as 'Ha-Ha-Ha-Ha', does so by interrupting a single expiration. Humans produce a burst of well-defined staccato vowels. Noisy chimpanzee laughter lacks clear harmonic structure and is produced one laugh note at a time for every expiration and inspiration. If an arrow to the right represents expiration, and an arrow to the left inspiration, the difference may be represented as follows:

Human: → ha → ha → ha → ha →
Chimpanzee: → ha → ← ah ← → ha → ← ah ←

Ape laughter has the sound and rhythm of a handsaw cutting wood.

Provine considers the inability to modulate with precision the expiratory air flow to be at least as limiting to human speech as is the structure of the vocal tract. Both ape and human laughter constitute a form of social communication. Laughter, therefore, may well represent another aspect which governed the evolution of articulate human speech, by providing both a respiratory control mechanism and further pleasure in communicating.

The Symbolic Capabilities of Bonobos

A number of fairly recent studies, in particular those conducted by Sue Savage-Rumbaugh, Duane Rumbaugh and S. Boysen (1980), have uncovered extensive language capabilities in bonobos: the linguistic capabilities of humanity's closest relative are equivalent to those of a very young child. Bonobos have no trouble understanding spoken words and following fairly complex instructions involving novel situations. It is clear that they understand the words spoken to them, because they are able to pick out of a sizeable assortment of photographs (not previously seen) the picture depicting the word (potato, cat, ball etc.). Obviously, their vocabulary is, as with a small child, limited. However, given a large specially designed computer keyboard in which each key represents a single word, they are not only able to commit to memory several dozen symbols, but to use them to great effect to communicate with their human companions.

Two things are significant about these observations. First, it is important to note that the computer's keys are marked in abstract ways; that is, the keys for potato or for ball do *not* have on them pictures which in any way resemble a potato or a ball. Therefore there is no difference, in principle, between a bonobo 'reading' the keyboard and a scholar reading a pictographic or ideographic manuscript. Both contain abstract symbols, one for each word. Second, and further proof of the first: bonobos, who obviously enjoy communicating with their human playmates, are able to create novel ***sentences.***

These discoveries should put aside, once and for all, the anthropocentric speculations of certain linguistic philosophers. Instead, we should heed the pronouncements of Charles Darwin, who, over a century ago, stressed that the difference between us and the rest of the animal kingdom is one of degree, not of kind.

The Trouble With Consonants

Consonants give shape to words: they provide a backbone delimiting the vowels. When they are lacking, one can only express utterances like 'oo!', or other emotional outbursts similar to animal calling systems. As an experiment, the reader should try talking using only vowels. This will demonstrate the profound limitations of such a linguistic system. Such utterances are exactly the sort of sounds used by almost all animals, particularly monkeys and apes, whose higher intelligence allows them to use this means of vocal communication.

To go beyond that, one needs consonants. This is the problem confronting bonobos and other apes: they are unable, or at least are limited in their ability, to pronounce consonants. Their vocal apparatus was designed only for vowels. This may also have been the problem for those species of *Homo* which became displaced by modern *Homo sapiens sapiens.*

Still, today, children have trouble with consonants. We shall discuss in Chapter 10 that perhaps up to 20 per cent of children fail to develop normal speech. A major form of phonological processing deficit involves problems with consonants, such as not being able to discern the difference between 'ba' and 'da', or between 'fa' and 'va' (Merzenich et al., 1996). Normally, such consonants each take only ten milliseconds of speech time. Once these sounds were stretched to 40 to 80 milliseconds, the language-learning impaired children had little difficulty in distinguishing 'ba' from 'da'. Incidentally, these language-learning impairments have nothing to do with general intelligence. Children who score very highly on other tests trying to assess intelligence may still suffer from a consonant-processing deficit.

It is not unreasonable to postulate that there was a time, earlier in the evolution of *Homo sapiens*, when the vast majority of individuals would suffer from what, today, we would class as phonological processing deficit – in particular, deficits which relate to articulating and discerning consonants. Such earlier species would not lack the intelligence to acquire speech, but would not yet have achieved the phonological sophistication needed for modern speech.

There exists a small category of words, the ***onomatopoetic words***, in which the word itself imitates the particular sound represented by the word. Examples of this category include 'buzz', 'hiss', 'gong', 'gurgle' and 'hum'. The attempt to imitate some of the sounds of nature might have provided for early speakers both an incentive for and practice in using certain consonants.

Other Changes

In order to generate consonants with ease, the anatomy of the human vocal apparatus had to become modified. As Lieberman has emphasized: 'So critical is the overall configuration of the vocal tract... that only when the voice box has retreated into the throat is the capacity for speech virtually certain'

Not only did the larynx move down into the throat, but also the tongue moved back while the jaw shortened and lost some teeth. The mouth, therefore, became less efficient for eating, and better adapted for talking – another indicator that nature was putting a premium on speech.

While the mouth and throat were evolving into an ever more efficient vocal apparatus, the brain, likewise, was evolving. In mammals the throat and mouth muscles are controlled by a part of the brain called the 'frontal cortex'. In apes, the motor area which controls these muscles is centred towards the rear of the left side of the lateral frontal cortex, approximately where ***Broca's area*** is located in the human brain. Broca's area is named after the French neurologist who discovered it back in

1861: people with lesions in this area are unable to produce all the vowels and consonants which comprise human speech. Broca's area must have evolved into a refined speech controller from its earlier, primate function as a more crude controller of the throat and mouth musculature. Interestingly, damage to the frontal cortex of macaques – monkeys much more distantly related to us than are the apes – also causes them to lose their ability to communicate. One can only agree when Lieberman states that: '...the neural mechanisms associated with language have a counterpart in the brains of our predecessors'.

Vocal tracts, as Lieberman points out, do not leave a fossil record. However, the anatomy of the human speech apparatus is such that it requires a greater bend at the base of the skull. In contrast, '...a straighter, unflexed basicranium is the hallmark of a non-human vocal tract. This is true for the skull of Neanderthal man... indicating that... that hominid was incapable of human speech'. The oldest known skulls exhibiting a highly flexed basicranium were excavated from the 150 000-year old Broken Hill site in Zambia. This is consistent with the idea that human language, as we know it, originated in East Africa.

Language and Genetics

There is another line of evidence which indicates that language originated with modern *Homo sapiens* in East Africa, then spread as result of migrations. This is based on an examination of a large body of genetic data.

Before we review this material, a word of caution: at the time of writing, there is still a good deal of controversy about the precise interpretation of these data (see the reviews by Barinaga, 1992; Gibbons, 1992, 1993; Klein *et al.*, 1993). However, the evidence suggests that this modern species, *Homo sapiens sapiens*, which includes all of us, originated in East Africa around 200 000 years ago.

Some geneticists have suggested that all modern humans derived from a single female – a hypothetical 'Eve'. This hypothetical Eve did not live alone. She probably had brothers and sisters – certainly numerous cousins. But it was not their offspring who survived to migrate all over the globe. Parenthetically, one cannot help but marvel at the intuition of the author(s) of ***Genesis***, who had Eve's first-born son, Cain, after killing his brother Abel and being banished from Eden, migrate to the land of Nod, where he met his wife and conceived Enoch (Genesis IV:16,17). Our hypothetical Eve might be the genetic grandam of us all, but she did not live alone – and her offspring migrated and interbred with others.

Most scientists interpret the 'Eve' hypothesis as follows: that we derive from a human sub-population – a population of Adams and Eves, perhaps as many as five to ten thousand of them – which subsequently spread over the globe. In any case, there is a general agreement that these events took place in East Africa; that modern humans, all of us, carry the genes of our African ancestry.

There exists more than one line of genetic evidence for the above. One of the most important of these is based on an analysis of DNA found in certain organelles of the human cell – the *mitochondria*. Mitochondria are derived from the mother. An analysis of such mitochondrial DNA shows that Africans have a much greater diversity than the inhabitants of any other continent. This implies that people have lived in Africa longer, because it would take more time to accumulate the larger number of mutations in their mitochondrial DNA.

The conclusions obtained by studying the DNA of mitochondria are consistent with other studies by geneticists focusing on the DNA of the cell's *nucleus*. Nuclear DNA contains the primary blueprint which determines those of our individual traits – metabolic, physical, intellectual, psychological – that we have inherited. The geneticist Luigi Cavalli-Sforza (1991) analysed the vast knowledge which had accumulated over a 50-year period about human genetics. This information included studies of over 100 different inherited traits which had made their appearance in 1800 different populations around the world – from the deserts of the Kalahari to the frozen tundras of northern Canada.

By looking at the frequency of the appearance of certain mutations in the nuclear DNA – for example, the occurrence of the Rh-negative factor in the blood – he was able to confirm (wholly independently) the African origin of modern *Homo sapiens*. Furthermore, his data showed that there occurred a number of migrations out of Africa, as well as across Eurasia and into Australia, the Pacific and America.

Other lines of evidence come from the fossil record. Fully modern humans appeared earlier in sub-Saharan Africa than in Europe (Brooks, 1988a,b). According to Christopher Stringer (1988) of the Natural History Museum in London, modern humans are of African origin. It was not until some time after these modern humans had spread around the globe that they differentiated into the various diverse groups and races which make up humanity today. (At present, not all palaeoanthropologists agree with this.)

Finally, and of greatest interest to us here, is the origin and evolution of human speech. At the time of writing, the sequence of events appears to be as follows: about 100 000 years ago, in East Africa, human speech was sufficiently well developed to be a recognizable language called *proto-World*. This was the original human mother tongue (Shevoroshkin, 1990).

These early modern humans, speaking *proto-World*, began to migrate. As they migrated and settled down, *proto-World* began to be spoken in various dialects. In due course, these dialects evolved to become separate languages. The first split involved the *Khoisan-speaking* people – the Hottentots, Bushmen and other tribes of southern Africa – the oldest linguistic phylum of them all. A second split resulted in *Congo-Saharan*, accounting for the peoples of central and northern Africa.

Apparently, by 35 000 years ago a derivative of *proto-World* had evolved among the settlers of the Middle East which, in turn, gave rise to a series of other proto-languages: *Nostratic*, *Sino-Caucasian*, *Austric*, *Indo-Pacific* and *Amerind*.

Nostratic existed 14 000 years ago in the Middle East. Over the next 4000 years it gave rise to the next set of languages: locally, to *Afro-Asiatic*, while migrants to the north came to speak *Uralic*, to the Far East *Altaic*, and, covering a broad band from Europe to India, *Indo-European*.

Afro-Asiatic evolved into *Semitic*, the precursor of *Hebrew* and *Arabic*. *Proto-Indo-European* appears to have been well established by 5000 years ago on the Eurasian steppes east of the Dnieper River in what is now the Ukraine and Russia (Anthony, 1996). The speakers of *Proto-Indo-European* lived at a time when the earliest Neolithic economies were established in this temperate zone. Migrations, isolation, conquests and assimilation caused a fragmentation of the language, so that by 1500 BCE three of its daughter languages – Greek, Hittite and Indic – had become quite dissimilar.

Shevoroshkin describes the further fate: by 1000 BCE the *Indo-European* family had fragmented into five major language groups in Europe (*Baltic*, *Slavic*, *Italic*, *Teutonic* and *Celtic*) and in Persia and India into *Iranian* and *Indic*. These latter two, in turn, gave rise to about 20 languages. Over the past 2000 years, the Teutonic tribes of Europe's North and West fragmented further linguistically, into Gothic (extinct), German, Danish, Norwegian, Swedish, Icelandic, Dutch and English. Similar events took place in other parts of the world, accounting for the ethnic diversity of the many thousands of peoples and languages found on this planet.

For example, *Amerind* was spoken by the earliest migrants from Asia into America, some 12 000 years ago. A later wave, about 9000 years ago, spoke a *Sino-Caucasian* derivative called *Nadene*. Last to arrive, about 6000 years ago, were the Eskimo-Aleutians, who spoke a language akin to

Altaic. *Altaic*, itself a derivative of that ancient Middle East proto-language *Nostratic*, was also the progenitor of Japanese, Korean and Turkic.

The Importance of Speech in Human Evolution

The acquisition of speech made us human. It did so in three ways. First, it greatly improved the capacity to communicate between individuals and groups. Second, and much more subtly, it changed the way we think and respond. Third, it became the basis for an oral tradition which provided a much more effective collective memory.

The depth of the impact of human speech can be fathomed by looking at its principal effect: it allowed communication across generations, so that a store of collective wisdom could be built up much more rapidly than by other means (such as learning by imitation). Instead of starting all over again at the beginning of every generation, each new generation could assimilate the experience and wisdom gathered over many, many generations. The emergence of speech as a form of social technology – once established – must have had a high selective premium, extending the *social* intelligence of the hominids into an ever-increasingly effective *collective* intelligence. **Cultural evolution now became more important than biological evolution** – to use the terminology of the biologist Richard Dawkins, the *meme* superseded the *gene* (Dawkins, 1976). (The meme is the fundamental unit of culture which is able to mutate, just as can the gene, the fundamental unit of genetics.)

The idea that languages influence the way we think will be discussed shortly in relation to creating a uniquely human internal information environment. It is likely that the great increase in brain size of the later hominids represented the high selective value placed on the ability to think in abstract terms and communicate more effectively. The increase in brain size is one of the more dramatic examples of a biological change brought about by the acquisition of a major new technology – human language. (The development of bipedalism, in response to the increasing reliance on the forelimbs to use weapons, is another.) After the introduction of speech, selection favoured those of our ancestors who could learn and remember words, and then articulate phrases and sentences.

It was a time when the poet ascended over the brute.

It was also a time when grandmothers became important. Grandmothers – post-reproductive females – have long been a biological anomaly: how can 'survival of the fittest' work when grandmothers are no longer able to transmit their genes to the next generation? Nor are grandmothers much good at protecting the young from predators.

The answer lies in the fact that natural selection may operate at *many* levels of organization. In the case of the late hominids, the individual had become organized into tightly knit social groups. The chances of the individual surviving and passing on his or her genes to subsequent generations depended on the success of the group. This is where grandmothers came in. They were the repositories of group experience and wisdom. Unlike grandfathers – who were more likely to get killed during the hunt or while protecting the group from predators – grandmothers were more likely to reach a ripe old age, especially if they were relieved of their reproductive functions. Grandmothers now had a job much more important than bearing children: they had become humanity's first specialized information storage and retrieval system.

The anthropologist Chris Stringer states this idea more cautiously. He also links it to a second idea: the long time it takes human infants to reach maturity. Both may be 'linked to the importance of intergenerational transmission of cultural information' (Stringer, 1988, p. 269).

Compared to other species, juvenile chimpanzees mature into adulthood only very slowly. But it is nothing compared to the human species. It was not always so. Newer techniques for analysing fossil remains to determine the age at death of the individual indicate that the modern human growth pattern is a recent development. It appears to have been lacking in the earlier species of *Homo* (*H. habilis*, *H. erectus*), probably also in archaic *Homo sapiens*, and perhaps even in Neanderthals.

This slowing down of maturation, of keeping things in a juvenile state for longer periods, is another example of 'neoteny' – a trick used by nature to achieve significant changes in structure or function. It has occurred at least three times in the biological evolution of human beings: first, in the case of our naked bodies: our skin never matures enough to become truly hirsute; second, our brain keeps growing for a longer period of time, probably because its embryonic growth shut-off has been postponed (Bonner, 1980); and third, as just discussed, overall maturation is retarded to allow for a longer period of learning. The evolution of the second and third of these features was crucial for learning language.

Homo sapiens sapiens invaded Europe and displaced Neanderthal man (*Homo sapiens neanderthalensis*) – not a stupid and insensitive creature by any means, but whose more limited linguistic capabilities, perhaps even limited to screams, hoots and grunts, were no match for the articulated communications of his modern cousins. In fact, whether human language originated with one brainy Eve or 5000 of them, she, or they, gave rise to a population of humans which, over the next ten or twenty thousand years spread all over the globe, interbreeding with or displacing earlier species of *Homo* – no doubt chattering all the way.

Human speech must have set the stage for the more effective domestication of plants and animals, one of the most profound of technologies – the basis of the Neolithic Revolution – which could ensure a family a steady, year-round food supply. Colin Renfrew (1989), Professor of Archaeology at Cambridge University, who believes that the Indo-European language originated in Anatolia (present-day central Turkey) rather than on the Eurasian steppes, has argued that the spread of the Indo-European language from Anatolia about eight and a half thousand years ago was associated with the spread of farming: farming tribes could have increased their population density 50-fold over the one person per 10 square kilometres characteristic of hunter-gatherers.

Summary and Conclusion

Human speech probably evolved from a combination of three quite different kinds of pre-existing primate disposition: (1) individual vocalization of the type observed in vervet monkeys, (2) group vocalization as exhibited by howler monkeys and carnivores such as wolves and coyotes, and laughter, as exists among apes, and (3) body language, particularly that which involves miming.

The emergence of speech as a property of human society illustrates a point which was made earlier. Just as a baby's babbling precedes the utterance of meaningful sounds, so is it probable that nonsense syllables shouted in cadence or sung while dancing preceded the emergence of meaningful 'spoken' sounds. When one looks at the evolution of semiotic interactions, almost always the message appears first with little or no meaning attached to it. Only later does it become codified and have attached to it a specific meaning.

The acquisition of linguistic skills would confer an enormous advantage on those who practised them. Once a truly human language arose, the original, now verbal, population would fan out to cover the globe, displacing their less articulate cousins. The evolution of the original East-African language, ***proto-World***, into the huge number of contemporary linguistic groupings, languages and dialects illustrates how a new information system may branch out in many directions as it, itself, becomes subjected to variation and selection. The evolution of language exemplifies the evolution of information-processing systems.

Once human beings had acquired language, a whole new dimension was introduced into both the human individual and the collective intelligence. The appearance of human speech resulted in a major discontinuity in the evolution of intelligence on this planet. Henceforth, ***cultural evolution*** would supersede ***biological evolution***. It is this phenomenon which differentiates us from the rest of the animal kingdom.

A similar discontinuity occurred in the evolution of ***meaning***. As the information transfer systems changed from ***zoosemiotic*** to ***homosemiotic*** systems, there arose not only a new message system – language – but also a new internal information environment, created by ***language***. As the ability to ***speak*** expanded, so did the ability to ***interpret*** what was spoken. The new information-processing systems evolved whole new layers of incoming-message/brain-context interactions, permitting ever higher levels of abstraction. Conversely, as the internal information environment expanded so would it favour an increasingly complex output. Language became more and more elaborate and messages achieved a deeper and deeper meaning.

The replacement of a sensory internal information environment by a linguistic one has been responsible for leading philosophers astray. In humans, attaching meaning to incoming information involves primarily linguistically based thought processes. This is not true for animals, which have not developed such linguistic competence. Furthermore, the discrimination carried out when certain kinds of information are selected out of an environment filled with information – and hence have meaning for the recipient information system – is a broad, universal phenomenon which, like intelligence, has undergone a long history of evolution. This universal phenomenon arose when self-organizing systems began to discriminate and utilize those elements in their surroundings which would aid in the survival or reproduction of the system. Thus, to be understood, the concept of meaning must be viewed as a spectrum of phenomena whose most primitive aspect is already exhibited by nonliving systems, while its most advanced form appeared when human language evolved.

· 10 ·

The Internal Information Environment

Introduction

When not ingesting, digesting or excreting, what an infant spends most of its time doing is tooling up its brain. What it is attempting to create is a useful ***internal information environment***. Such an internal information environment becomes useful when it has created an accurate mental 'map' of the external world – a map which allows the infant to interact successfully with all aspects of its environment and manipulate it to its own advantage.

The Internal Information Environment

In addition, this information environment must be organized in such a way that the flood of new information which impinges on the baby's brain through its sensors may be assigned ***meaning***. The brain must learn how to filter out information of no significance to the organism, retain information which has significance for the moment or for a short period of time, and store information which may be helpful at some point in the future.

The information stored in the brain must be organized in such a manner as to facilitate information retrieval. This means that there must be a system for organizing the stored information. The matter is analogous to the technique used by another information storage system: libraries. For example, libraries may use the Dewey decimal system to assign new books to their appropriate shelves. In like manner, so must the brain evolve a system for storing new information inputs in a way which allows easy retrieval. The brain, however, has one major advantage over the library: unlike the library, which usually can afford to store only a single copy of a book, the brain may store the same information at numerous sites, creating webs of interrelated (contextual) information.

Jeremy Campbell, in his book *Grammatical Man: Information, Entropy, Language, and Life*, uses the phrase 'a syntax of memory' (Campbell, 1982, pp. 214–15) to describe how information is assimilated by the human brain. Campbell recapitulates what has been known for a long time: whereas disorderly information which lacks structure is easy to forget, information which is in an orderly state, rich in pattern and structure, highly interconnected and containing a good deal of redundancy is much easier to remember – '...new information is remembered better if it is embedded in a wider context of information already known from personal experience'. This latter, of course, is what constitutes an important part of the internal information environment.

As the brain matures, new information coming in finds more and more ready-made niches to fit into – each such niche providing a 'context' for that new piece of information. Thus the maturing brain creates more and more opportunities for the incoming information to become meaningful. In turn, the more meaning can be attached to a new piece of information, the more easily it is remembered. In this manner, the brain develops an ever more complex internal information environment

Organizing the day's information inputs probably occurs during sleep (Evans, personal communication). The reason babies spend so much time sleeping is not merely because their muscles are tired but because they need time out from further external inputs to integrate the newly acquired information into their existing store of knowledge. It is this store which becomes the ***major*** part of the individual's internal information environment.

There is another part of the internal information environment which must not be overlooked: the 'hard-wired' neural networks. These are responsible for more primitive forms of behaviour, such as the muscular movements associated with locomotion – movements whose proper co-ordination must be learned but for which there exists a neurological ***anlage*** (predisposition). A second, more complex, neurological *anlage* is manifested by the numerous instances of complex behavioural responses observed throughout the upper end of the animal kingdom which do not appear to be learned, and which therefore must be classed as ***instinctual***. In human infants this appears to include a fear of the dark and a fear of falling.

A third category of biologically based behavioural responses, based on hard-wired neurosystems found in the brain, are those relating to the phenomenon of ***imprinting*** (Slukin, 1965). One of the earliest observations involved incubator-hatched male chaffinches kept in soundproof isolation. Such birds begin to sing their personal, if incorrect, version of the species' mating song during the first spring after hatching. If they hear a correct version of this song, either from a live male or from a recording, they quickly learn the correct song. However, this can occur only during the first spring. If they are not able to modulate their own version and correct it during that period, they are never able to learn it; they are doomed to loneliness for all succeeding springs (Thorpe, 1961).

Human children are able to learn to speak foreign languages without a trace of accent. During their mid-teens, this capacity is usually lost. If a person learns to speak English as an adult, no matter how erudite a scholar, no matter how marvellous a command of the English language – extensive vocabulary, even perfect grammar – let us say an Albert Einstein – the accent always provides a clue to the native English-speaking listener of the person's foreign birth. In fact, acute listeners are able to distinguish readily among the various foreign accents whether the speaker comes from Europe, India or Japan, or, for that matter, from which part of Europe.

The imprinting phenomenon is the perfect example of how the internal information environment is created by a combination of the neurological *anlage* and the external information inputs. It is the existence of such neurological predispositions which aids the exploring infant in creating an extensive, and truly human, internal information environment. Specifically, the verbal inputs, including those coming from a child's own voice-box, resonate in the speech centres of the brain. These centres have been genetically selected to facilitate the most efficient way of creating a powerful internal information environment, an environment which is able to process and assimilate with incredible speed the information presented by verbal inputs. One of the main functions of the genetic hard-wiring must be to help discriminate between the subtle sounds emitted by the human vocal cords which make up a child's social environment. Crucial to the acquisition of language is to learn the nuances of both strings of phonetic sounds making up words and strings of words making up phrases and sentences.

Recent work (see Tallal *et al.*, 1996) has shown that infant language acquisition begins during its first month of life. At that time, exposure to a specific language begins to shape an infant's phonetic perception, resulting in the acquisition of a set of ***prototypic phonetic representations***. These become the building blocks out of which the maturing child constructs its spoken native tongue.

Normally, this process occurs without any explicit instructions. However, it has been estimated that about 20 per cent of children fail to develop normal speech. About two-thirds of these are the result of sensory or cognitive deprivation. It is becoming clear that ***phonological processing deficits*** – the inability to discriminate between the various sounds making up a language – if not corrected during infancy, are likely to result in serious language-learning impairments at a later stage in life, including reading disabilities.

As a child matures, an incredibly rich ***linguistic*** internal information environment develops. By late teenage, in the USA a High School graduate is able to recognize on average 80 000 words (Miller and Gildea, 1987). Assuming that all of this word learning took place over a sixteen year period, the child learned between infancy and adulthood close to one hundred new words each week. Children with large vocabularies probably pick up new words at twice that rate. This feat is not achieved by memorizing word lists, but simply by the maturing child encountering words in some context of interest and relevance, hence ***meaningfulness.***

Ascribing a meaning to a word involves at least two stages. The first is a rapid one in which children are quick to assign a new word to a category of things: i.e. into ***semantic*** categories. Even though they may not be able to remember the specific name of the new word, when they do encounter it again they know what kind of a word it is. As Miller and Gildea point out, children who may not remember the name of a new colour, and therefore respond with another (if incorrect) colour name, would not respond with terms such as 'round' or 'five' or 'lunch'.

The second stage involves sharpening up the distinction between words within a semantic category: e.g. how 'olive' differs from 'brown' or 'green'. This process takes place over a longer period of time.

The creation of such a rich linguistic internal information environment allows a person, when confronted by holes of information – as for example, when one has a conversation at a noisy party and misses a third of what has been said – still to arrive at the intended meaning of the conversant partner. Even nonsense words may yield a meaning, as is illustrated by the first stanza of a famous poem:

> 'Twas brillig, and the slithy toves
> Did gyre and gimble in the wabe:
> All mimsy were the borogoves,
> And the mome raths outgrabe.

Lewis Carroll, Jabberwocky, in Through the Looking-Glass[1]

---

1 For an interesting analysis of this poem, including Carroll's own set of definitions, see Gardner (1960).

Note that of the 23 words which comprise this stanza only the following eleven are common words in the English language: 'the' (used four times), 'and' (used thrice), 'did', 'in', 'all', and 'were' (used once each). 'Twas' and 'gyre' are rare and are used mainly for poetic emphasis. Yet this short piece, part of the most famous of nonsense poems in the English language, conjures up images and a mood – images and moods very much dependent on the internal information environment brought to the poem by the individual reader.

It should become obvious from the above that the internal information environment acquired by the human brain through learning is profoundly different from that of most animals. The internal information environment acquired by animal brains through learning consists of interconnected ***sensory*** memories – sights, sounds, smells, tastes and tactile experiences, as well as any other physical phenomenon which the animal is capable of detecting (e.g. magnetic or electric fields). In contrast, the human brain has, in addition, a huge store of abstract information based on ***linguistic*** experiences. (The same may prove to be the case for certain species of dolphins.)

In humans, the linguistically based experiences are probably the dominant form of information stored in the brain. However, in spite of our conceit to the contrary, the sensory is of great importance as well. This is easily demonstrated by reflecting on the fact that in our dreams, lengthy sequences may occur as event follows event, and image follows image, without a single word being uttered. Or consider the *déjà vu* experience when a subtle odour, not consciously perceived, causes our stimulated internal information environment to transport us back to another time. Our brains, as with the brains of other animals, are able to record and think in terms of sequences and relations of sensory experiences alone. It is by utilizing such internal networks of sensory inputs that it becomes possible to create a 'map' of the world – a map which does not require words.

However, what distinguishes us from the rest of the primates is that even our sensory experiences become translated into linguistic ones as the brain processes and analyses the sensory information. When we engage in heavy duty thinking it is not generally in terms of unlabelled images, sounds, smells, tastes or tactile experiences. Rather, it is something which Charles S. Peirce described best when he posited that: 'thinking is talking to oneself'. This ability to talk to oneself is so basic a part of our human internal information environment that it tends to shape all our thought processes. It is this fact that allows us to be so influenced by our social and cultural surroundings.

On Misunderstanding

The above analysis provides us with a clearer picture of how messages become misunderstood. There are two quite different mechanisms by which misunderstanding may occur. The first involves the message itself. It may become compromised during transmission. An example of this is illustrated by the story of Julius Caesar sending a message to save one of his soldiers from execution: 'Spare, not execute!'. Unfortunately, the message arrived with the comma transposed to read: 'Spare not, execute!'. To avoid this type of misunderstanding, languages have built into them a great deal of redundancy. Caesar's message with proper wording – 'Do not execute but spare!' – would have been more difficult to compromise. It is this type of message corruption that has been the primary concern of communications engineers like Shannon.

The second involves the difference in the internal information environments of the sender and of the receiver. No two human beings have the same internal information environment. First, there are differences in the neurological hardware provided by each individual's genetic heritage. Second, each person's life experience is different: earlier, we considered the differences in the perception of the four individuals confronted with a piece of Johnny's jigsaw puzzle. Third, part of that life experience is the cultural heritage which shapes the individual's perception and competencies. A rather trivial example to illustrate this is the reaction to the statement: 'I'll knock you up in the morning'. Uttered by a male acquaintance, this is likely to elicit a response of anger or fear from an American female. For her, this slang expression implies a sexual liaison in the morning leading to pregnancy. In the north of England, however, this expression may merely signify a knock on the door to awaken you in the morning. From a Yorkshire lass, therefore, placing the message in context, it is likely to elicit a friendly response of thanks.

The Mind as a Form of Virtual Reality

Virtual reality is based on a highly sophisticated merger of sensor technology, computers and output devices. The sensors detect a person's movements, including the direction in which the eyes are moving, then transmit this information to a computer for analysis. The computer responds by activating visual images, sound and pressure devices to simulate an artificial environment.

The impact of virtual reality on a person has been admirably described by Steve Aukstakalnis and David Blatner (1992) in a book aptly entitled *Silicon Mirages*. As the authors point out, even though the scenes are mere simulations, and even though they may still be quite crude, they exercise a profound emotional impact on the person entering such a world: '...something astounding happens... you are in two places at once: your body is encumbered with gadgetry, but your mind and much of your ***perceived*** reality is hovering around somewhere else'.

The authors go on to point out that in the early days of film, 'film-naïve' audiences reported feeling that a train rushing towards them on the screen was real. This illusion did not last long for regular film buffs: the screen was, after all, not three-dimensional, did not surround the viewer and was not interactive. Virtual reality has managed to correct all three of these deficiencies, even though this technology is still in its infancy.

Virtual reality requires active physical responses of the participant to perceived situations. This involves activating motor neurons: passive learning becomes transformed to experiential learning. One result of such sophistication is that the intensity of the ***virtual*** experience may be such that it becomes recorded by the brain as a genuine ***physical*** experience. It becomes increasingly difficult for the brain to distinguish reality from simulation.

One reason for our inability to distinguish fact from fiction when confronted by *virtual* reality is that the brain recalls all experiences by ***simulating*** them (if imperfectly). When virtual reality impinges on the senses in a manner sufficiently close to physical reality, the brain records them as real experiences. Only flaws in the virtual environment, providing clues to its 'unreality', and the silent commentary from somewhere within the cortex that the experience is not really real, prevent the brain from becoming completely fooled.

As discussed earlier, information is stored as ***patterns*** of neural connections. The brain is able to recreate these patterns. When it does so, it ***simulates*** previous experiences of the brain. Simulating a previous experience constitutes a memory. Any number of things may trigger off any particular memory: a simple sensory input such as the smell of lilacs, a more complex sensory input such as riding a horse, a single symbolic input such as hearing a simple command, or a more complex symbolic input such as reading a book.

In addition, we may talk to ourselves inside our heads using words, or run a series of images in a stream of consciousness – all of which triggers off neural discharges along one or more specific pathways – pathways predetermined by the brain's earlier experiences. In the process, it simulates these earlier experiences (real or imagined) – creating further memories. On top of that, there is a whole host of accidental firing of neurons, whether awake or asleep.

Once started, any neurological activity in the brain tends to start off other pathways which may lead to other images and thoughts. Under normal circumstances, the brain never shuts down completely or stops after a single image or thought.

Memory traces need not involve pathways identical to the original. Undoubtedly, the greater the differences from the original the fuzzier the memory. This is where Edelman's neural Darwinism comes in (see Rosenfield, 1988): the pathways are being continuously revamped according to usage; hence the old pathways are being continuously corrupted. The less used they are, the faster they become corrupted.

There are certain exceptions. The evolution of the brain has created special categories of nerve traces which become relatively incorruptible. For example, those created during the developing embryo relating to vital functions are probably stable during our entire lifetime. Those involving imprinting phenomena in juveniles, likewise, are less prone to decay. The same thing may be said for those which become habits and those learned during emotional stress.

Given the complexity of the cortex – ten billion (10×10^9) cells and 60 trillion (60×10^{12}) synapses (Shepherd, 1990, p. 7) – the brain can store and recreate an almost unlimited array of images. Some of these images recreated by the brain may be 'real' in the sense that they are genuine recollections produced by activating the appropriate memory traces. Such mental images help us recognize and interpret visual data. Others may be purely imaginative, involving novel combinations of neural pathways. These occur regularly while dreaming, and may become troublesome while hallucinating under the influence of drugs, a high fever, brain damage or other pathological conditions.

Like the computer, therefore, the brain also creates a kind of virtual reality, reflecting a basic dualism: the physical world outside and the neurological representation of that world inside. It is this dualism which has preoccupied philosophers for millennia – it is a controversy which has raged in philosophical circles for centuries. Up until the time of the appearance of computers, it was almost impossible to resolve. Now, because we have machines capable of carrying out thought processes, there exist models which allow us a clearer conceptualization of the problem.

Our experience with computers has taught us the difference between 'hardware' and 'software'. Hardware is the computer itself. Software is the program stored in the computer. Hardware is analogous to the brain; software is analogous to the mind. In both the computer and the brain information exists in the form of *patterns*. The software resides in the computer as a *pattern* of on/off switches; the mind resides in the brain as *patterns* of neural connections. As neural network computers come on stream they will look and act more and more like the human brain. Von Neumann computers need to be programmed by an external intelligence. Neural network computers, like the human brain, learn by being exposed to external sources of information.

In spite of the differences between the neurological representations of the brain and the electronic representations of the computer it is not beyond the realm of future possibility that there may arise a technology capable of transducing one form into another. However, this is a matter to be explored in a future work. What is important here is to recognize that the internal information environment provided by the human brain is one which simulates reality – that the semantic space in which words conjure up images is, in fact, operating within a form of virtual reality.

On the Nature of Understanding

For the literal minded, the term 'meaning' may not be applicable when it comes to the response an ant makes to a piece of sugar. At best, the sugar may have 'significance', but the ant cannot comprehend what 'sugar' means in the way humans can. Therefore, it cannot be meaningful.

This is a dangerous intellectual quicksand. Unless one argues that the term 'meaning' has applicability only to human beings, one cannot in logic argue that an ant lacks the intelligence to pick out a grain of sugar from all its surroundings. The intellectual quagmire is resolved by understanding that 'meaning' is a phenomenon which represents a spectrum – a spectrum which evolved alongside both the increasing complexity of information-processing systems and its product, the increasing sophistication of intelligence.

Operationally, whether or not a piece of information is 'meaningful' to a recipient system may be determined by observing the actions taken by the system following the reception of such information. In the case of proto-intelligent systems, such as crystals, enzymes and antibodies, such actions consist of carrying out the appropriate chemical or crystallographic processes for which the system was designed.

In intelligent systems, meaningful information results, normally, in behaviour which is intelligent, i.e. behaviour which enhances survival or reproduction, or which in some other way fosters the achievement of some goal. (This is not always the case: under certain circumstances, the combination of message and internal information environment may lead to counterproductive behaviour, indicating that the system is not beyond making mistakes.)

This conceptualization of 'meaning' overlaps with the definition of 'information' used by library information scientists. As cited earlier, Brian and Alina Vickery (1992, p. 30) consider that information: '...is any input into the system that initiates a change of state'. This statement appears to be in general agreement with the thoughts expressed in the present work. However, the term 'a change of state' as used by Vickery may also apply to non-intelligent systems, where it becomes equated to 'a change in organizational state', while in intelligent systems it leads to 'a change in behaviour'.

Since proto-intelligent systems are capable of extracting meaning from information in their environment, it must become apparent that intelligence is ***not*** a prerequisite for ascertaining the meaning of incoming information. Which brings us to the important distinction between *meaning* and *understanding*! Only higher animals are capable of understanding. Proto-intelligent systems and intelligent systems lacking a nervous system merely ***react*** to environmental stimuli without understanding the significance of their own actions. Ants represent a step up, although it seems unlikely that an ant has much understanding of what is going on in its environment. In higher animals with advanced brains, one gains the impression that not only are they able to understand situations in a much broader context, but that they have attained an integration of environmental information inputs such that they may be said to possess ***consciousness***. As was discussed in Chapter 4, higher apes exhibit an even more advanced internal information environment in that they have a demonstrable sense of ***self***, i.e. they possess ***self-consciousness***.

Not only apes, but also certain species of dolphins have a sense of self-identity. This is based on a number of observations, including the fact that each dolphin, among its many vocal signals, has a signature call. In the case of dolphins, their mode of perceiving their environment by means of echo location and how they communicate this to each other suggests the possibility that, in addition to a single conscious self-identity, they may also have a communal identity. Such an identity is denied us humans, for whom consciousness is a purely private affair (Jerison, 1986).

The identification of self is probably more widespread among higher animals than we are currently led to believe. As was pointed out in Chapter 4, higher animals which live in societies where there is both a lot of communication and a lot of competition need to become 'natural psychologists'. Such skills, which involve understanding the moods, thoughts and intentions of conspecifics, would greatly aid survival in mutually interdependent social groups. As Donald Griffin (1984a,b) has pointed out, it would seem inevitable that the daily comparison between others' and self states would, sooner or later, lead to a sense of personal identity.

A sense of self is a necessary condition for that ethereal phenomenon we call ***consciousness***. However, it is not a sufficient condition. Consciousness requires that we are also aware of ourselves as knowers. We will have a brief look at consciousness shortly. The point here is that we have stumbled into the morass of an evolutionary spectrum where it becomes extremely difficult to define the demarcations in any sharp and clear manner. 'Understanding', like meaning and intelligence, involves a gradation of phenomena. A fish in a tank, responding to the environmental clue of a person who is about to feed it, exhibits a primitive level of understanding of the situation. Much more advanced are the ingenious efforts of a squirrel trying to feed from a bird feeder which has been 'squirrel-proofed'. We are led to believe that human understanding is at the top of the scale, but, as we saw earlier in the example of the jigsaw puzzle, even here such understanding involves gradations depending on the nature of the message and the cultural context in which the message is interpreted.

Understanding goes beyond meaning. Meaning, as stated above, involves the integration of a message into the internal information environment of the recipient. Such a process creates a new information unit: the combination of the external information complexed with the information provided by the internal information environment. This unit will be referred to from here on in as a 'semantic complex'. Such a ***semantic complex*** may be further information-processed as if it were a new message in its own right. By repeating this process, the original message becomes more and more meaningful as, at each recursive step, new semantic complexes are created. As these impinge on ever larger areas provided by the internal information environment, whole new and elaborate knowledge structures may be built up – a process which leads to ***understanding***.

Thus, in the human brain there are created elaborate knowledge structures. These involve external inputs which have been recursively information-processed. Such information processing leads to hierarchies of information inside the brain. Perhaps one measure of meaningfulness is the amount of information processing stimulated by the incoming message as the new information is incorporated into existing knowledge structures. It is as if a rolling snowball creates an avalanche. The recursive information processing of new (meaningful) complexes of information may, at some point, result in creating a new and novel knowledge structure. Semioticians would refer to such a phenomenon as 'emergence', physical scientists would call it a 'phase shift', and the rest of us would recognize the event as the transformation of information into ***insight*** and ***understanding***. We shall return to this matter in the next chapter, as we consider the idea that a piece of information entering the human brain is cycled and recycled through several existing knowledge structures in a manner suggesting a ***semantic metabolism***.

Consciousness

Consciousness is a phenomenon not yet clearly defined. Professor Aaron Sloman of Birmingham University considers the concept of consciousness to be a 'cluster concept'. Such a cluster consists of a large number of capabilities which can be present or absent in different combinations. To quote Sloman (1996, p. 44): 'Instead of asking which animals are conscious and which are not, we might, for instance, talk about 75 different kinds of consciousness and ask which animals have this kind, which have that kind and when do we have this kind and which kind do we lose when we have Alzheimer's disease and which subset do we have when we are very young and so on'.

In addition to the fuzziness of the concept in Western scientific circles, there is an almost unbridgeable schism between the Western and Eastern traditions. The former tries to define consciousness in terms of mental processes with particular emphasis on the chemical and neurological mechanisms underlying brain functions. The Eastern tradition focuses more on the various states of consciousness and their effect on the individual. The focus of the former is rational, the latter spiritual. One is reminded of Rudyard Kipling's dictum: 'East is East, and West is West, and never the twain shall meet'. Some psychologists, such as Allan Combs (1995, p. 168), have tried to start an 'honourable dialogue' between the two. Combs views consciousness as a system near the edge of chaos, and states of consciousness as chaotic attractors which draw the system into patterns of activity, or basins.

The Eastern tradition is beyond the scope of this book. However, what both traditions agree upon is that there exist different states of consciousness. In the Western tradition an alert, task-oriented state is different from a daydreaming state, or one in deep meditation. A brain at sleep and dreaming differs from that of an individual in a hypnotic trance, and various trances, hallucinations and drug-induced states of consciousness all differ.

How these different states relate to meaning is almost trivialized by the statement: changes in the conscious state of an individual change the internal information environment of that individual. This means that the same (external) information will interact quite differently with a recipient who has undergone a change in consciousness state – as if it were encountering a different mind. A classic case is the interaction between two individuals who, when sober, meant very little to each other, but under the influence of alcohol found each other rather attractive, and hence had acquired a new meaning for each other. Likewise, people who have experienced drug-induced states of consciousness know how the same object can look very different while on a trip.

If consciousness cannot be achieved until a brain has evolved sufficiently to allow for such a phenomenon (phenomena?) to occur, then on the basis of all that has been said earlier, meaning and consciousness are quite separate and distinct occurrences. Meaning does not require intelligence to manifest itself. Consciousness does.

However, if meaning is a phenomenon which does not require consciousness, the reverse is not true. Consciousness involves a whole series of meaningful information inputs which are processed further in the internal information environment of the mind. It is this processing, soon to be described as ***the semantic metabolism*** of advanced brains, which allows this phenomenon to occur. The emphasis must be on the term 'advanced' brains, which implies two important aspects: first, less advanced brains are not capable of consciousness, or are capable only of what might be considered to be primitive forms of consciousness; and second, like all the other organs of the human body, brains, too, have an evolutionary history. Such a history implies that consciousness also evolved and probably did not arise as an all-or-nothing property of the growing intelligence.

In light of the above, it seems reasonable to conclude that, just as both intelligence and meaning involve a spectrum of phenomena, so does consciousness.

The idea that consciousness involves a ***spectrum*** is applicable not only from the point of view of phylogeny, i.e. its evolution, but also from the point of view of its ontogeny, i.e. as the infant grows into a child and then matures into adulthood. This latter is a theme admirably developed by Susan Greenfield in her book *Journey to the Centers of the Mind* in which she considers (Greenfield, 1995, Preface, p. x) that: 'Consciousness is not all-or-none but increases and deepens as the brain becomes more sophisticated and progresses from fetus to neonate to child'. Particularly important appears to be a shift in cognitive capabilities when the human infant reaches about 18 months of age (Meltzoff, 1990, p. 25). As psychologist Andrew Meltzoff describes it, there is a shift: '...from using empirical or experience-based representations to using hypothetical representations, which concern possible realities. This development shift allows children to project into the future... and deduce from past... in advance of, and sometimes in the absence of, strictly perceptual evidence'.

In adults, brain damage may cause a reversion to an infantile state of consciousness. Patients with right brain damage were uncertain as to what was important in a story and what was incidental (see Campbell, 1982). They seemed unable to guess future events in the narrative, even when these events were quite predictable. Such patients had a good grasp of syntax and spelling and the meaning of the individual words and phrases, but had trouble with placing the information into a context. A good deal of recent work on brains indicates a substantial difference in the kinds of mental activities which predominate in each brain hemisphere. A rough description of the difference is that the left brain acts more like a smart Von Neumann computer, conversant with logical deductions, while the right side retains much of our hominid emotional makeup, good at picking up patterns of information and reacting to it intuitively.

All of this illustrates that it is probably better to talk of many different types of consciousness, rather than to lump the whole thing into one catch-all concept. Or if it is treated as a unitary concept, it would be more appropriate to treat it as a cluster concept, such as 'health'.

Meaning as Applied to the Internal Information Environment of Cells

The question remains: how does the human brain process a message to derive meaning from it? Instead of looking to computers for a useful model, it might be that a better paradigm is furnished by the internal information environment of ***cells***. If the brain represents the hardware, and the mind the product of that organ functioning, then the organelles, membranes, structural proteins, enzymes and other constituents which give a cell its structure are its hardware, while its dynamic metabolism represents the product of that cell functioning.

This dynamic metabolism is subjected to a large number of regulators – messages which originate both from within and from without the cell.

The most important messages emanating from within the cell are those sent out by the DNA. We have touched upon the role of DNA in regulating cell metabolism in Chapter 6. Suffice it here to indicate that the language of DNA is universal among all earthly biosystems known. The actual messages emanating from the DNA will, of course, vary from organism to organism, and from cell to cell in a multicellular organism. Most important for our discussion here is the theme developed by Lila Gatlin a quarter of a century ago (Gatlin, 1972, pp. 132–41). Various codons (consisting of three nucleotides each) and anticodons (consisting of three nucleotides which are the 'mirror' image of their respective codons) may be interpreted differently by the cell depending on the context in which they appear. Thus the codon AUG (adenine–uracil–guanine) means 'START' when it appears at an initiating site, and means 'methionine' (an amino acid) when it appears in the interior of a polypeptide message.

The above is an example of 'meaning' at the cell's most fundamental level – its biochemistry. The following represents an example of how two different kinds of cell exposed to the same chemical messenger may respond to its message very differently, depending on the internal information environment of the recipient.

Well over a score of years ago, Gordon Tomkins (1975) reviewed the phenomenon of 'biological symbolism' and the origin of communication between cells. In most microorganisms, the accumulation of cyclic AMP[2] in the external environment acts as a signal for carbon source starvation. In like manner, ppGpp (guanosine 5'-diphosphate 3'-diphosphate) represents a signal signifying nitrogen or amino acid deficiency. Note how these chemicals bear no structural relationship to the molecules whose shortage leads to a nutritional or metabolic crisis. Cyclic AMP is not a chemical analogue of glucose, so therefore it is an *abstract* signal and must be considered as a *symbolic representation* of a feature of the external world.

As with any symbolic message, its meaning depends on the internal information environment of the recipient. The bacterium *Escherichia coli* responds to the build up of cyclic AMP by synthesizing the protein flagellin. The production of flagella (whip-like organelles which propel a bacterium) is a response to a shortage of utilizable carbon source in the immediate environment. Cyclic AMP therefore acts as a message which is interpreted as: 'food is running short, let's move'. As a result, a colony

---

2 The reader unfamiliar with biochemical terms should not worry about the chemical nature of cyclic AMP (also sometimes further abbreviated as c-AMP). What is important is that this molecule seems to have been designated by nature as a ***messenger molecule***.

of bacteria disaggregates as its individual cells become motile and disperse.

Quite opposite is the behavioural response to cyclic AMP by the slime mould *Dictyostelium discoidium*. When the nutrient-supply status of the is favourable, this organism exists in the form of dispersed independent cells which resemble amoebae. As conditions become unfavourable, these 'myxamoebae' generate cyclic AMP and release it into the surrounding medium, where it acts as a chemical attractant: the individual and dispersed amoebic cells aggregate to form an organized structure which matures into a fruiting body. This fruiting body produces spores able to survive the adverse times.

The two organisms handle the impending food shortage quite differently. However, in both instances, the chemical message furnished by the environment is the same: an excess of *cyclic AMP*. How the message is interpreted and what response follows depends on the internal information environment of the respective organisms. In each case, this internal information environment provides a ***context*** which allows the message to be interpreted. To emphasize the point once more: the 'meaning' of the message depends on the decoding apparatus available. For *E. coli* the message decoded to result in behaviour which led to a ***dispersion*** of the cells; in *D. discoidium* the decoding resulted in ***aggregation.***

Bacteria and other single-celled organisms are constantly bombarded by messages advising them of a change in the state of the environment (see Kaiser, 1996). This is equally true for cells of higher organisms. One example of how the interpretation of a message depends on the internal information environment of a cell is illustrated by the taste buds of the tongue. In cells which are responsive to sweet taste sugar boosts the level of cyclic AMP, while in a bitter-responsive cell sweet things are ignored but a bitter substance like quinine causes cyclic AMP levels to be depressed (see Barinaga, 1996). In general, in higher organisms, the role of cyclic AMP is to act as an interpreter of the primary message. This is particularly true for hormones, which are considered to be the primary messengers. Cyclic AMP may therefore be classed as a second messenger, although there are several intermediaries between a hormone and the cyclic AMP.

The release of cyclic AMP leads to a cascade of regulatory and metabolic events. The initial information, provided by the environment in the form of a hormone, becomes transduced and cycled and recycled through the metabolic machinery of the cell. This metabolic machinery represents, of course, the internal information environment provided by the cell. This interrelation of events leads to the idea that, perhaps, the brain likewise engages in its internal information environment in something which might be called ***semantic metabolism***.

Let us explore this matter further.

· 11 ·

The Semantic Metabolism

Introduction

The idea of semantic metabolism is this: when there is an information input into the human brain, such as a visual observation or an auditory message, the information is metabolized by the brain the way a molecule of glucose or an amino acid is metabolized by the cell, or the way a hormonal message entering the cell is cycled throughout the various cellular systems. Since the cell itself is an immensely complex information-processing system, it could serve as a useful model for the information processing carried out by the brain.

Having said that, there are substantial differences between a cell and the brain. The analogy should not be carried too far. However, the currently popular 'the brain is a computer' paradigm is also much too simplistic, and the concept of a semantic metabolism should provide a useful adjunct.

Cells receive information from their environment all the time – information which is decoded by putting it into a chemical, metabolic or physiological *context*. Such a process can take place only because the cells provide an internal environment which allows them to respond to external chemical stimuli in a highly selective manner.

A precise description of the myriad feedback control systems which comprise the metabolic machinery of cells has become one of the hottest topics in biological research. The past few decades have seen amazing progress in spite of the almost unfathomable complexity of cellular metabolism. Among other things, there has been a shift in emphasis from trying to integrate the diverse metabolic activities of the cell by applying 'energy flow' as the means of integrating this diversity to recognizing the cell as a complex information processing system: it has been estimated that the development of a human being is guided by just 750 megabytes of digital information stored as DNA molecules in the egg (Olson, 1995). By using the flows of information, what could otherwise be a system so complex as to escape understanding becomes conceptually integrated.

It is not the purpose of the present work to review these complex cellular information-processing systems.[1] However, there are two points which need to be made. First, the unravelling of the complexity of a cell's information-processing system is perhaps as difficult an endeavour as, in the future, it will be to tease apart the processes by which the human brain processes information. Second, since the study of cellular metabolism is so much more advanced than that of the equivalent work on brain information processing, might the cell system not serve as a model? Information theorists will need to address themselves to the following question: in principle, how similar is the information processing carried out by cells and that carried out by the brain?

To explore this possibility, we need to look briefly at some of the metabolic events which occur within cells.

Hormone Messages and Cellular Metabolism

Hormones, the 'chemical messengers' of the human body, may be divided into two major groups based on their solubility in water (see Bolander, 1995). Those that are relatively insoluble in water – the so-called hydrophobic ligands – have no problem in crossing the cell membrane and migrating into the cell's nucleus, where they may activate or repress individual genes or gene complexes, exercising a profound effect on the cell's long-term metabolism. The second group are water soluble and do not traverse the cell membrane. Instead, these hormones – the 'hydrophilic' hormones – interact with receptors: molecules distributed over the surface of the cell.

Receptor molecules may be thought of as long protein molecules. At one end, where they protrude outside the cell, they complex with the hormone. At the other end, which protrudes into the cell, they are capable of initiating a series of metabolic reactions. In other words, the receptors are membrane proteins which transmit the message (represented by the hormone) from outside to the inside of a cell. Four basic super-families of such receptors are recognized at present. It is not necessary, for our

1 The reader is referred to the paper by Artavanis-Tsakonas *et al.* and the other reviews appearing in the 14 April 1995 issue of *Science*, an issue devoted to ***signal transduction.***

purposes, to go into further detail (see Bolander, 1995) except to point out that the receptors are able to transduce (convert) external information in a matter analogous to the human eye or ear. In the human case, light or sound is converted into nerve impulses. In the case of the cell, the chemical message (e.g. a hormone) is converted into a cascade of metabolic events.

Once the message has entered the cell's interior, its internal information environment takes over. Just as nerves from the retina, or those from the inner ear, reach into the depth of the human brain, so do receptor molecules manage to transmit information into the depth of the cell's metabolism. For example, receptors may activate so-called ***G proteins***. These proteins act as molecular switches: ***Gs*** stimulates the enzyme ***adenylate cyclase***; ***Gi*** inhibits it. Adenylate cyclase is the enzyme which synthesizes ***cyclic AMP*** from ATP. Cyclic AMP, in turn, is able to activate several enzymes and ion channels; in particular it can activate ***protein kinase A***. Protein kinase A is one of a class of enzymes called 'kinases' because they change the kinetics of reaction of other enzymes. Protein kinase A activates other kinases. These, in turn, activate other enzymes or still more kinases.

Activating different kinases in series, ultimately leading to the activation of the target enzymes, is the chemical equivalent of an electronic amplifier system: a weak incoming signal is amplified so as to make the information decipherable to the recipient. For a cell this amplification is crucial, because hormones are present in the bloodstream at concentrations of only parts per billion (10^9) while their receptors rarely exceed more than a few thousand per cell.

The cell is a much more complex system than an electronic circuit. It should come as no surprise therefore that, in addition to amplifying the incoming message, several different kinds of kinase may participate in modulating the various metabolic reactions. This modulation involves cross-talk between various systems and kinases; it creates what Pedro Marijuán of the University of Zaragoza has called 'the society of enzymes' (Marijuán, 1994). This ability to communicate so effectively results in the exquisite coordination of a cell's metabolism and its stability.

In addition to activating protein kinase A, cyclic AMP can have other effects. In particular, it can act on another superfamily of receptors – the ***ion channel receptors***. Ion channel receptors act like pores in a cell's membrane which allow inorganic ions to enter or leave. Both cyclic AMP and cyclic GMP cause these pores to open and allow cations (positively charged ions) to migrate across the membrane. Among such cations is the calcium ion. Elevated levels of calcium inside the cell fluid can bring numerous cell changes, affecting enzymes, the cytoskeleton itself and other physiological processes.

Cyclic AMP is considered to be a 'second' messenger which carries out the instructions furnished by the hormone (which has remained outside the cell). Besides cyclic AMP, there are numerous other second messengers which activate all kinds of other systems in a cell. There are other systems in which a water-soluble hormone which reacts with a receptor initiates a series of events which include sending a message (in this case a so-called zinc finger protein) to the nucleus (Galcheva-Gargova *et al.*, 1996).

After a thousand million years of evolution the cells of multicellular organisms have devised numerous means to achieve whatever ends needed to be attained. Among such ends was the need to coordinate the behaviour of cells in a multicellular context. Intercellular communication was achieved simply by coupling external chemical signals to the pre-existing intracellular regulators, which then became a set of second messengers. As we have seen in the previous chapter, in some cases the reverse also seems to have occurred: a well-established internal messenger such as cyclic AMP would be secreted into the external medium to signal other cells about the impending shortage of carbon sources.

The Concept of Semantic Metabolism

At this point in history it is a truism to say that the brain 'processes information'. This particular terminology derives from our recent historical experience with computers – devices which are able to carry out mental processes, that is, processes which previously could be done only in our heads.

Ideas about how the brain works have always been influenced by what technology was at hand. During the 20th century various suggestions have included: (1) that the brain functions like a complex series of electronic circuits; (2) that the brain is a telephone exchange; (3) that the brain is a kind of library; and (4) that the brain functions by creating holograms. Not only new technology but also new ideas originating in other fields have also been applied. One of these is chaos theory and its supplement, complexity theory. We touched on this briefly in the previous chapter, when discussing Allan Combs' view of consciousness.

All of the ideas present above have some merit to them. There are probably aspects of brain functioning which resemble the workings of electronic circuits, telephone exchanges, libraries, holograms and chaotic behaviour, including attractors. However, the dominant paradigm over the last few decades has been that of the ***computer***.

The matter may be compared to the state of ignorance a few centuries ago, when the functioning of another mysterious organ, the heart, became understood. This understanding came only after Europe began to experience the extensive use of pumps following the Middle Ages. The ancient Greeks, having no experience with pressure pumps, considered the heart to be a hearth. This was not a bad supposition, for when the heart stopped beating, the blood grew cold.

It is in this manner that we have become accustomed to the idea that the brain engages in information processing. What advantage, then, might be gained by saying that the brain engages in metabolizing ideas; that we are not dealing merely with information processing but, more specifically with a semantic metabolism? The answer has several parts.

First, the term 'information processing', as we have discussed in Chapter 3, must apply to many different systems. Some of these systems have very little to do with how the brain functions. The term 'information processing' is, therefore, too general and gives us no idea how the brain actually carries out such processing.

Second, and almost opposite to the first, is the fact that when the term 'information processing' is used, generally people do not think of the many forms of information processing which exist, but rather of the processing carried out by computers. This means that the phrase is too specific: it misleads people into thinking that the brain thinks the way a computer does. This is a fallacy. Computers are logic machines; brains are not, or only secondarily so.[2] As discussed earlier, this is beginning to change with the advent of neural network computers.

What the term 'semantic metabolism' is trying to achieve is to create a paradigm shift. How the brain functions should not be modelled only on computers, even neural network computers, but also on the information processing carried out by cells. Even if, at this point, we cannot fill in many of the details, there are some real advantages to conceptualizing how the thoughts traversing the brain undergo transformations in a manner which parallels the way information is handled by the cell.

The brain is a complex organ. The computer is not. The cell is intermediate between these two. That is not to say that advanced computers, capable of engaging in massive parallel processing, are not complex devices. However, they do not, as yet, approach the sophistication of the

2 A comparison of classical computers with the human brain is presented in Appendix D.

information processing carried out by cells. For one thing, cells are able to reproduce themselves; computers cannot.[3]

There was a time, a mere generation ago, when the matter of unravelling the mysteries of the cell seemed almost hopeless. The rapid strides made in understanding the complexities of the cell as an information-processing system should put to rest any pessimism that the brain is too complex a system, a system whose mysteries we shall never unravel.

Most Cells can Act as Both Conduits and Modulators of Extra-Cellular Information

In earlier sections we have considered how a chemical signal, such as cyclic AMP or a hormone, leads to a cascade of metabolic events which change the behaviour of the recipient cell. One response to an incoming message is that the recipient cell, in turn, sends out messages to influence the behaviour of other cells. The cells of the brain and nervous systems – the neurons – are specialized for carrying out this function. This specialized function for the analysis and transmission of information makes a neuron a natural intermediary between the information processing carried out by the more typical cell and the information processing carried out by the brain. It may be that the neuron's specialized function will serve as a better model for the semantic metabolism carried out by the brain.

On the other hand, many other cell types also engage in acting as conduits and modulators of information. The cells of various glands of the human endocrine system, when stimulated by hormones released by the 'master gland' – the pituitary – will secrete their own hormones, such as thyroxin, adrenaline and oestrogen. The endocrine system is, of course, a master coordinator of our body's physiology. Many other cells of multicellular organisms perform similar functions, creating chemical feedback loops, particularly during the growth and differentiation of the embryo. Usually such cells have other functions as well; in fact, such cells may engage in message exchange only as part of a secondary function. The best examples may be found in plants, which lack a nervous system altogether but whose growth is as regulated as that of animals, and in some plants, such as the ***Venus flytrap*** and the ***Touch-me-not***, one may observe the equivalent of nervous activity.

---

3 Although, in theory, they could create automated computer factories to create more of themselves – see Appendix C.

The occurrence of cells which act as both conduits and modulators of information in multi-celled organisms is a frequently encountered phenomenon. It may even prove universal. In most instances, the message is not merely transmitted unchanged; rather, as in the example of the endocrine glands cited above, the message put out by the recipient cells is different from the message coming in from the pituitary. The gland's output will have been modulated by a series of external and internal inputs.

Discussing the endocrine system moves our focus from the cellular level to that of the entire body. As discussed in Chapter 3, the human body is an information processing system of extreme complexity, made up of many information processing ***sub***systems. When the concept that the brain ***metabolizes*** information reaches its full flowering, it may be that the metabolism of an entire organism will prove more relevant than that of individual cells. However, at this stage of development we shall use the metabolism of the cell as our guiding principle.

Parenthetically, research on cellular metabolism in general will yield dividends in understanding how neurons work. (To some extent, the reverse will also happen.) This should not be taken out of context: no amount of understanding of how individual neurons work will lead to an understanding of how the brain thinks, just as no amount of understanding how a transistor works will make comprehensible how a computer computes. However, the reverse is not true: it is impossible to have a complete understanding of either the brain or the computer without an understanding of how a neuron or a transistor works.

What is germane here is the question of whether the principles of general cellular metabolism will reflect general principles of information processing. The hope is that a highly complex system, capable of message transmission and modulation – the cell – will provide a paradigm for the semantic metabolism carried out by the human brain.

Some Comparisons Between Cellular and Semantic Metabolism

Let us begin our comparison between cellular metabolism and semantic metabolism by considering that two classes of substances enter a cell. The first class of substances involves those which can be classed as 'nutrients' – molecules which contribute to the growth of the cell; the second class is those classed as 'messages' – molecules which change the direction of the metabolism and the behaviour of the cell. In a like manner, we can differentiate between two kinds of information input in the human brain. The first, like nutrients, contributes to the growth of the knowledge structures accumulating in the brain; the second – messages – causes the brain to alter the behaviour of the individual.

One might argue that messages also contribute to the growth of the knowledge structures, but then some metabolic messages cause a quiescent cell to start growing again, or a dividing cell to accelerate its growth and reproduction, by supplying a vital growth factor. Thus in both cases we must admit that the divide between nutrients and messages is not a hard and fast one.

When it comes to messages, a second parallel phenomenon is observable. Hormones, as discussed above, comprise two major groups. Those in the first group, which are insoluble in water, traverse the cell membrane and the interior of the cell with ease, affecting the recipient's DNA directly. These hormones have a longer term effect, such as changing the reproductive behaviour of the cells or organisms of which they are a part. The second group of hormones, although their messages require receptors to enter the cell, initiate a cascade of metabolic events which cause a rapid response once the message has been transmitted to the cell's interior. This rapid response may bypass the cell's genetic apparatus.

This second type of hormone response corresponds to information inputs which generate reflex reactions. Reflex reactions bypass large areas of the brain in order to respond rapidly. They involve a response very different from information inputs, which are less alarming and which involve elaborate semantic pathways to generate more reflective reactions.

Substances such as hormones which regulate the growth and differentiation of cells do so by activating kinases which modulate various enzymes (including other kinases). G proteins, referred to earlier, may activate or inhibit adenylate cyclase, the enzyme which synthesizes cyclic AMP. All of these are examples of how a cell regulates the flows of information and its metabolism by ***modulating*** the activities of its multiplicity of enzymes. This leads to an obvious parallel in brain physiology: enzymes operating in a cell may become activated or inhibited by other enzymes just as neurons in the brain may become activated or inhibited by other neurons.

This principle extends well beyond individual enzymes or neurons. Whole areas of metabolic activity may be inhibited (or enhanced) by the cell, just as major neurological structures may be inhibited or activated by the brain.

As was discussed in Chapter 5, during the growth and development of multicellular organisms individual cells are subjected to different chemical gradients and combinations of gradients. These gradients, which lead to the activation and repression of genes, ultimately lead to a cell's specialized differentiated state. When compared with the original fertilized egg, each specialized cell represents a 'new idea'. In the brain, 'new ideas' frequently come about when two or more information inputs combine to create a new semantic complex. 'Inspiration' may be the result of gradients floating throughout the brain's semantic space intersecting in a manner to produce a new insight.

In cells there exist numerous metabolic cycles. In the brain, the idea that a piece of information entering the human brain is cycled and recycled through several existing knowledge structures is supported by various observations starting to accumulate on mechanisms of visual imagery and perception (see Miyashita, 1995). Mutual images stored, as it were, deep in the brain, can be recalled and made to interface with new perceptual representations coming in from the eyes at early stages of information processing. Such recalled images can actually interfere with the accuracy of the incoming visual data. For example, subjects were assigned a vernier acuity task, such as determining whether two line segments were perfectly aligned: when the subjects did not form mental images accuracy was 80 per cent, but when they formed images accuracy was only 65 per cent.

Professor Sloman, cited in the previous chapter, illustrates this type of phenomenon even more dramatically (Sloman, 1996, p. 41). The reader is invited to consider the following statement, part of a well-known adage:

A
BIRD
IN THE
THE HAND

to see if you can spot an error. Most people see nothing wrong with this statement, even after perusing it several times. It is only when they are asked a leading question that it becomes self-evident.[4] The incoming information, because of its familiarity, is rapidly metabolized by the brain – too rapidly – substituting memory for reality.

Professor Sloman considers that brain 'architecture dominates mechanism'. He discerns three layers in this architecture: (1) the automatic reactive processing layer; (2) the management layer, which constructs and contemplates options in advance; and (3) the metamanagement layer, which monitors, evaluates and redirects high internal actions.

4 How many times does the word 'THE' appear?

Each of these layers was added on during the course of evolution. A similar course must have occurred during the evolution of the cell. Simple prototypes evolved into prokaryotic cells, which evolved further into eukaryotic cells containing nuclei, which, in due course, added various cell organelles.

It is immaterial whether it turns out that the human brain contains three or several dozen layers. The complexity of the semantic metabolism evolved as the brain, during its evolution, acquired an increasingly complex architecture. The same holds true for the evolution of the metabolism of the cell.

Semantic Metabolism in Neural Networks

There is another reason for preferring the concept of ***semantic metabolism*** over ***information processing***. We trivialize the complexity of neural circuits when we compare them to electronic circuits. Each individual neuron has an astounding number of capabilities built into it. For example, it has the capacity to act as a conductor of impulses (like a wire, or optical fibre) and as an attenuator of signals (like a resistor), as an integrator of inputs (like a capacitor); it has its own, built-in power source (like a battery); and it acts as a gate for thousands of other neurons. In combination with other neurons it is able to create circuits whose subtlety and complexity is beyond the wildest dreams of contemporary electronics. Furthermore, as we shall discuss in the last chapter, the sheer number of elements making up the human brain endows it with information-processing abilities which are orders of magnitude greater than those of present-day computers.

As indicated in Chapter 7, if present-day computers are still relatively simple-minded (to coin a phrase) they will, at some point in the future, exceed human neurological capabilities. Part of such technological progress will come about as a result of the evolution of neural network computers. Already, today, one can point to examples of semantic metabolism carried out with neural network computers.

Hinton *et al.* (1993), using networks of simulated neurons programmed appropriately, have taught such a computer to learn how to read. Of great interest is the fact that, once the system has been trained to read, upon being deliberately damaged it will make the same types of bizarre errors when reading a work as will humans who have suffered certain types of brain damage. Such a network was able to reproduce both the 'visual' and the 'semantic' errors made by patients suffering from deep dyslexia. In the former, the error is based on the fact that the two words ***look*** alike; in the latter, the two words have related ***meaning***. Sometimes the computer, like the patient, committed both kinds of error. A patient referred to as G.R. would read 'sympathy' as 'orchestra', presumably via 'symphony'. The damaged network of the computer would read 'cat' as 'bed', presumably via 'cot'.

The authors conclude that the ***meaning*** of a word appears as if it were a point in a ***semantic space***. The region around each word represents what in chaos theory is referred to as a ***point attractor***. Once a network's state enters such a region, it will cause it to be inexorably drawn to the point represented by that word. Because such regions overlap, and because the semantic space is multidimensional, it becomes easy to see why an impaired system ends up in an adjacent region which has at its centre a point containing a word which looks similar (a visual relative) or has a similar definition (a semantic relative). It must be obvious from this that the internal information environment comprises not only *what* information is stored by the recipient, but also *how* it is accessed and retrieved.

Epicentres, Gestalts and Consciousness

The existence of the human brain, with all its complexity, can be accounted for in terms of classical evolutionary processes. However, given the advanced architecture of this brain, this does not explain how that architecture relates to its capacity to engage in a sophisticated semantic metabolism.

A child is born with very limited capacity. Although its brain is not a blank page, a '*tabula rasa*', the bulk of its semantic metabolic activity develops as the newborn matures into the child and beyond into the young adult. This increase in brain complexity occurs not by adding neurons but by vastly increasing the connections between them. This business of interconnection is the key to understanding how the brain works.

Initially, the complexification of the neural network must result from the newborn engaging in mapping the basic features of both the body and the environment. Although human babies are born with a strong grasping reflex, to coordinate all the movement of a human hand is a process which takes years. Learning to walk, and later to run, is another of those processes which teaches the network to fix a whole slew of interconnections. Distinguishing light from dark and the colour, shape and distance of objects requires much upgrading of the developing nervous system. And so forth.

Essentially, these are processes carried out by all mammalian young. In fact, to a greater or lesser degree they must be carried out by all animals which have a brain. Among social animals, the brain must not only learn to map its body and its physical environment, it must also map its social environment. As was discussed earlier, in animals with advanced brains such social interactions probably lead to a sense of self-identity. In addition to that, there is the question of that ill-defined phenomenon, ***consciousness***.

Susan Greenfield, in her ground-breaking book, *Journey to the Centers of the Mind* (Greenfield, 1995), considers that consciousness arises when a stimulus generates an epicentre among certain neurons, which then recruit neighbouring neurons into a massive, coordinated bundle of nerves. She refers to such a neural assembly as a ***gestalt***, which she defines as: 'a highly variable aggregation of neurons that is temporarily recruited around a triggering epicenter' (Greenfield, 1995, p. 112).

According to Greenfield, as children develop they have a gradually growing consciousness of the physical world around them. Colours, shapes and sounds – initially all jumbled up – slowly become converted to mother, food, window, bedclothes, crib.... In due course, mother and other members of the family, including pets, are differentiated from strangers. At a certain stage in an infant's development, strangers are viewed with much suspicion, if not downright fear. Later, this fear is overcome as the child learns to manipulate strangers. If the smile is an inborn feature of our humanness, when to apply it most effectively is learned. Thus, the developing child not only has a growing awareness of its physical surroundings, it becomes increasingly conscious of its social environment. All of this is associated with strengthening (or weakening) neural connections.

Each object, each new circumstance and each new idea generated inside the brain leads to an association of neurons – an epicentre – which grows into a neural assembly, i.e. a gestalt. Not all neural assemblies are gestalts, but all gestalts are neural assemblies. Spreading out from the epicentre, the interconnected cells work in unison. In some instances, the distances involved are sufficiently great that local neurotransmitters activating groups of adjacent or nearby cells are not likely to play a role. There is evidence that disparate neurons, at some distance from each other, can actually oscillate in a synchronous manner. Greenfield makes the analogy of 'synchronous swimming' and cites a number of other workers who observed that when the thalamus and the cortex are active in unison the animal engages in focused arousal. In addition, a similar synchrony may occur among subsets of neurons in the cortex, where individual neural members of such a gestalt can be recruited up to a distance of seven millimetres away.

Such gestalts are a clear example of collective intelligence and exhibit a property of such intelligence, namely ***synergy*** – the total is greater than the sum of its parts.

Greenfield provides numerous correlations between mental phenomena, including consciousness, and the presence, size and permanence of gestalts. Her arguments are persuasive. They represent a useful set of working hypotheses trying to correlate what goes on in the brain with what goes on in the mind. She is much to be commended for that.

There is a second theory, not so much in opposition as complementary, which emphasizes another aspect of consciousness. Stephen Kosslyn, Professor of Psychology at Harvard University, has proposed that consciousness is a kind of 'parity check' (Kosslyn and Koenig, 1992, pp. 434–7). This concept is based on a computer metaphor: computers store information as 'bytes', which are sets of 8 'bits'. Each bit is a binary digit ('0' or '1'), and seven of them could represent a set of 128 characters (enough to cover all the keys on a typewriter keyboard). The eighth bit is often utilized as a 'parity bit', which is set as '0' or '1' in such a way that the ***sum*** of the eight bits making up the byte always comes out to be an odd (or, if so desired, even) number. If the sum comes out incorrect, the computer will recognize that one of the bits in the byte has been compromised.

Kosslyn's theory considers that consciousness evolved to act as a kind of parity check. If the brain is functioning normally and there are no obvious problems, the various parts of the brain, including Greenfield's epicentres, are 'in tune' with each other. That is, the various parts set up resonances which Susan Greenfield describes as 'synchronous swimming' but which might better be likened to an electric power grid humming along with all its participating power stations operating at a synchronous frequency. For the brain, neural discharges in various areas which process the same stimulus seem to oscillate together at a frequency of 40 Hz. According to Parity Theory: '...consciousness has the same relation to these brain events as a chord does to individual notes that are played on a guitar. Consciousness arises from the interaction of the electromagnetic rhythms set up in individual brain loci, but cannot be reduced to the individual brain events any more than a chord can be reduced to individual notes'.

Kosslyn points out that when the brain is operating in harmony, we are not conscious of our mental activity. It is only when the neural states do not mesh properly, when the guitar is out of tune, that consciousness comes into play. Consciousness has a unique role which cannot be replaced by any individual brain process – it represents a synergism, the product of incredibly large numbers of units (neurons and epicentres) busy communicating with each other. Consciousness, as a super-brain phenomenon, acts as a check and as an organizer of the system. It is a kind of information manager.

One minor criticism: the coordinated oscillations of the epicentres and other interconnected brain areas are not primarily an ***electromagnetic*** phenomenon. Neurons operate by the ionic depolarization of membranes and by the release of chemicals across synapses. This is how they transmit and modify information.

The transmitted and modified information, as it creates resonances, begins to have a life of its own – a matter to be considered in the final chapter. Such a phenomenon is important for understanding how a ***physical*** structure, such as the brain, is able to achieve a ***non-physical*** information reality which constitutes the mind.

Both Greenfield's and Kosslyn's conceptualizations of consciousness would be compatible with the idea that the fate of the various kinds of information which comprise our mental life resembles very much how a ***cell*** metabolizes information. And it is the nature and extent of such metabolism which determines the ***meaning*** that the brain ascribes to incoming information.

Is it Possible to Quantify the Meaning of a Message?

Earlier on, we alluded to the possibility of being able to quantify the meaningfulness of a message. For example, would it be possible to measure precisely the difference in meaning that the jigsaw puzzle piece had for the four parties involved?

At the moment, the answer must be negative. We do not, as yet, know enough about how the brain works to be able to define, with sufficient precision, the internal information environment of a recipient. This is also true for the cell. For, in spite of the enormous strides alluded to in the previous sections, we still do not know enough about the metabolism of a cell to describe clearly all the aspects of its internal information environment. As to the brain, there are major questions, such as how is the incoming information distributed so as to match it up with its existing knowledge store? How does it know where to find the relevant information? Is there such a thing as an 'attractor', as postulated by Hinton *et al.*? If so, what is the mechanism of attraction? And when this information (contained by the message) does locate the relevant internal information, what is the nature of the match-up?

These are some of the problems of today. As we have seen, it is not unreasonable that over the next few decades the advances being made in brain science, neurobiology, neural network computing and theoretical biology will provide the insights necessary for resolving these puzzles.

In the meantime, we should not sit by idly, contemplating our philosophical omphalos. Shannon and related information theorists have provided a mathematical analysis of the amount of information which may be ***transported*** within a given channel. This has proved most useful for solving various communications engineering problems and for assessing syntactical aspects of language. Is it not possible to devise some method for quantifying the semantic aspects of a message in the way that the communications engineers measure the relative size of a communication?

To restate the obvious, in order for an input signal to have 'meaning', the incoming message must be placed into a ***context*** by the receiver. Such a context is provided by the 'internal information environment' of the receiver. Thus the key question becomes: how is information processed (metabolized) to derive 'meaning' from it? This is as true for a single cell as it is for the human brain. This leads to a number of possibilities, some of which are discussed in Appendix E.

Undoubtedly other possibilities exist besides those described in Appendix E. The point we need to make here is two-fold: first, the efforts to quantify the semantic content of a message are premature – we simply do not know enough; and second, this will not always be so – as we work out the precise nature of the internal information environment provided by the human brain, we will come up with methods of quantifying the interactions which confer meaning on a message.

Summary and Conclusion

The human brain metabolizes incoming information. Its architecture, based on the massive interconnections of hundreds of thousands of millions of neurons (10^{11}), allows it to become so sophisticated in its information processing capabilities that it is able to create a semantic space which we call consciousness. This semantic space has no physical reality, but is a virtual phenomenon. Just as the information stored and processed by the computer is not made up of transistors and other physical components of the computer, so are the thoughts and information processed in our heads not made up of neurons and the other physical components of our brain. The computer's information derives from ***patterns*** of on/off switches; the brain's information derives from ***patterns*** of neural connections. Both create a virtual reality, a semantic space, which is 'real' but made up of pure information, not of matter and energy.

Consciousness involves creating hierarchies of information which commonly are described as 'data', 'information', 'knowledge', 'insight' and 'wisdom'. As was done in Chapter 1, this ordering of information may be compared with the ordering of matter when going from fleece, to yarn, to cloth, to clothing. In this context, the term 'information' is defined as 'ordered data', a conceptualization which is probably still the dominant one among both the lay public and many experts.

As shown earlier, to develop a coherent theory of information and meaning, we must recognize that, of the five categories, ***all*** represent information. Starting with 'data', which is, after all, merely incoming information, this incoming information is cycled and recycled through the brain; that is, it is metabolized. During this semantic metabolism it becomes increasingly organized as it is added to already pre-existing knowledge structures. It is the resultant more developed structures which may create an insight. Insights, when coupled together, result in wisdom. It is as if a cell, upon ingesting a variety of amino acids, links these into polypeptide chains. Polypeptides, in turn, become incorporated into a protein strand, and two or more such strands combine to form a complete protein. The protein may then enter into association with other proteins, nucleic acids, polysaccharides, membranes and other constituents to form a cell organelle... and so forth. It is no more miraculous that the patterns of neural connections existing in the brain give rise to 'the mind', than that simple organic molecules give rise to 'life'.

When we relate this process to ***meaning***, it would seem that the greater the semantic metabolic activity generated by an incoming piece of information, the more meaningful it is. In general, this is probably true, but only if the information is assimilated into pre-existing knowledge structures and contributes to building up the forms of information described above – knowledge, insight and wisdom.

· 12 ·

Review and Further Discussion About Consciousness

Introduction

Information which acquires new meaning for us can be likened to the flowering of a beautiful tree in spring. When we start to understand a foreign language, or to understand the meaning of a poem, the achievement of making previous information acquire a new or deeper meaning – the achievement of obtaining an ***insight*** – can be so profound and so joyous an event that we are left to wonder about the process which led us to such a state of grace.

Thus we begin to wonder about the meaning of meaning. In our thoughts, we come to recognize how unique is such an act – an act which can stir in us such a profound emotional response. Intuitively, we ascribe the response as being unique to ourselves and to other human beings.

And therein lies the fallacy. For what we are witnessing is the upper end of a spectrum of phenomena in which a recipient of information is able to discriminate among items of incoming information and merge such information with the information stored in the recipient's own knowledge structure. It is a phenomenon which has a long evolutionary history behind it.

In order to understand the concept of meaning in a ***human*** context, it is necessary to understand the concept of meaning in a ***biological*** context. And in order to understand the concept of meaning in a biological context, it becomes necessary to understand the context of meaning in a ***cosmic*** context. By cosmic is meant the entirety of the known physical universe: it is not meant to have any theological implications.

It is the intent of this present work, as part of a trilogy, to help lay the foundation for a science of information. Specifically, this tome is designed to review and expand on two earlier works dealing, respectively, with the physical basis of information and the evolution of intelligence, and then to go on to explore the phenomenon of meaning. For this reason, the book is divided into three parts, each dealing primarily with information, intelligence and meaning. Out of this synthesis there emerges a theory of how the brain is able to create a conscious mind – a matter to be covered in this and the next chapter.

Information

Information should be viewed in the same way as we view energy: as an intuitive first principle. Like energy, information should be considered as a basic property of our universe. Operationally, energy is defined as the capacity to perform work – not the work itself. In like manner, information should be defined operationally as the capacity to generate organization – not necessarily the organization, or the act of organizing a system, itself.

It is important to distinguish between information and information ***processing***. The term 'information processing' is frequently associated with computers. This is not surprising, since it is, in fact, what computers do. In the present work, however, 'information processing' is a term which is applied in its broadest sense – it considers 'information' to be the raw material upon which such processes work. Information processing includes the transduction, transport, storage and retrieval of information, as well as its analysis. As we shall review in the next section, information-processing systems have evolved into some remarkable forms.

Information and energy are readily interconvertible: the performance of work – which consists of a ***transient state*** – invariably leads to a reorganization of the universe (and hence its information status) – which consists of a ***stable state***. Physical scientists have been able to ignore the information component, which is present in all physical interactions, either by considering time, distance and physical constants as a universal given or by inventing accounting devices such as potential energy, latent heat and entropy.

Information manifests itself as patterns of organization: that is, any system which exhibits some form of organization contains information. This is as true for a pattern of molecules in a crystal as for a pattern of letters in a book. At the physical level, Boltzmann's statistical mechanical concept of entropy represents a quantity which measures the information content of a physical system, although it does it awkwardly: information represents a function of the multiplicative inverse of the exponent $e^{S/k}$, where S is the entropy.

This means that entropy is a physical measure of the ***absence*** of information: an increase of one entropy unit in a system represents a decrease in its information content of approximately 10^{23} bits/mole. It is unfortunate that the term 'entropy' has also been used as a metaphor by communications engineers and most information theorists. Their use, or abuse, of the term 'entropy' has led to no end of confusion.

Whenever 'work' is being performed, the physical organization of the universe is altered. Hence information and energy are constantly being interconverted, one form into the other. Furthermore, information, like energy, may exist in a particulate state, a form which we have referred to as 'infons' – virtual particles, like migrating holes in a crystal, which possess neither mass nor momentum.

Information Processing

As indicated above, the addition of information to a system results in greater organization of that system. The ability of information to organize itself, as well as matter and energy, results in the formation of ***information hierarchies.*** The complexity (from quarks to human cultures) which characterizes our earthly habitat – physical, biological, social, intellectual – reflects this property of information: its ability to organize not only matter and energy, but itself as well. As we move up from one level of complexity to the next, we observe the equivalent of phase transitions. Each higher level relies on new categories of bonds, or novel ways of using pre-existing bonds, to generate new types of organized structure. New types of organized structure allow for the evolution of novel information-processing systems.

In Chapter 3 we examined a number of information-processing systems. Information processing occurs inside crystals, among enzymes, within cells, among tissues and organs, among body systems such as the immune system or systems regulating the body's fluids, across nerves, in the brain, between social animals and, of course, in humans – ranging from a couple to human global society. In addition, information processing also occurs in mechanical systems, such as the home thermostat, and electronic systems, such as the computer. The evolution of information processing at the cosmic, biological, human and technological level has led to the many diverse ways of handling information: in the nature of the information itself, in the carrier which transmits such information from one part of the system to another, and in the manner in which it is stored. These systems are so diverse as to mislead information theorists. We should be no more astonished at this diversity of information-processing systems than at the fact that biosystems evolved both mice and elephants, hummingbirds and ostriches, snails and giant squid, mosses and redwoods, and bacteria and us.

The evolution of the universe may best be understood in terms of the increases in the complexity of information systems and the increased capacity of such systems to engage in information processing. Plotting the relationship between entropy and information results in a graph which appears to trace the evolution of the universe from a state of infinite entropy and zero information to the emergence of self-organizing systems, followed by self-reproducing systems, the formation of life itself, the increasing complexity of life forms, the origin of the human species, the rise of the collective intelligence of that species to include linguistic and other cultural systems, and, most recently, the emergence of machine and electronic intelligence.

To reiterate what was already emphasized in Chapter 3: information processing is going on all around us and within us. It organizes the universe, it sustains life, it gives shape to culture, and it is the basis of our thoughts – that is, our intelligence.

Intelligence

Intelligence has been defined as a property of advanced information-processing systems. It is a phenomenon that has arisen after a prolonged period of evolution. It is a state or a condition which can be ascertained only by observing the system in action: that is, whether it engages in intelligent ***behaviour***. Intelligent behaviour involves an ability to analyse the system's environment and then to make an intelligent response. A response, to be classed as intelligent, must lead to one of three outcomes:

1. The system has enhanced its own ***survivability***.
2. The system has enhanced its own ***reproducibility***.
3. If the system is goal-oriented, it has enhanced the achievement of that goal.

By this definition it becomes apparent that intelligence must be viewed as a spectrum of phenomena, ranging from the 'proto-intelligence' exhibited by pre-biotic and certain electronic systems to the most advanced form known to us: the collective intelligence of human global society.

Intelligence should not be confused with ***consciousness***, which is a property manifested only by advanced intelligent systems. When the brain circuitry gets complex enough to generate patterns internally – divorced from external stimuli – it becomes possible to engage in *introspection* and *abstraction*. Internally generated patterns, in turn, lead to the possibility of becoming aware of what one is doing now and of what one may want to do instead, and then to plan doing it. These are the roots of *consciousness*. When a mental map of the world includes the mental states of conspecifics, the stage has been set for *self-consciousness*. Understanding, even if only in a vague and imprecise manner, the mental states of others must also generate an image of *self*. The recognition of the mental states of other beings in animals which exhibit an extensive social organization must be vital to their reproductive success, if not their actual survival. Judging from their behaviour, such higher, social animals recognize many different kinds of message: 'this animal may threaten me'; 'this animal may mate with me'; 'this animal is warning me of some danger'; or 'this animal may cooperate with me to fight the challenger'. It is probable that the internal maps of all higher social primates have non-verbal representations of each individual member of the troop with whom they have had interactions. Such non-verbal representations would not only pertain to the physical characteristics of each individual, but also their social and psychological attributes.

At the moment, we are still reluctant to grant higher cognitive and intellectual capabilities to other animals. When the whole story is in, it seems likely that we will be amazed at the mental powers of higher animals – just as we are learning how much more extensive is the mental life of the youngest of human infants.

Just how Advanced is Human Self-Consciousness?

The phenomena of introspection, of thinking in abstract terms, of self-consciousness, is deeply developed in humans. However, in our anthropocentric egotism we overlook the fact that we humans also do not possess a self-consciousness which tells us what is really going on. *Cogito ergo sum* – I think, therefore I am – Descartes' axiom, is grandiose in its conception. However, it masks our ignorance. For although we *know* that we think, we do not know *how* we think. Our mental map of how our brain functions is largely a blank. We *know* that we remember, but we do not know *how* we remember.

Likewise, what drives us – our emotions – is about as clearly understood as was our physiology in the Middle Ages. Only now are we beginning to discover our evolutionary predispositions as a result of studying our fellow primates and, more generally, as a result of ethological studies which are helping us discover the laws governing animal behaviour.

Similarly, we may have a sense of self-identity, a self-image based on reflections in a mirror or the responses of family, friends and workmates, and perhaps even a public image based on reports in the mass media – however, our self-image is usually fuzzy and out of date. This is especially true as we get older: 'I don't feel any different from when I was seventeen'.

At this point in the evolution of humanity, we probably could not face up to a model of self which is completely accurate. For one thing, it would have to include our senescence and demise – a singularly depressing matter, best left in mental limbo until it becomes inescapable. We also have a tendency for self-righteousness – we tend to think of ourselves as 'good people', particularly when we are in a conflict with others.

Our mental map of ourselves, of which we are so proud, is filled with uncertainties and delusions. As humanity evolves, one of the future improvements in our intelligence will be a better understanding of what it means to be human. There will come a point when we will truly understand ourselves as individuals, will understand our fellow human beings, and will understand our collective selves, that is, human society. This will come about as advances in the various sciences provide us with a better understanding of what motivates us, how our brain functions, how societies evolve and where we fit into the cosmic scheme of things.

Meaning

At the moment, there exists a major source of confusion in the field of information science: 'information' is confused with 'meaning'. A large school of thought does not consider information to be information unless it has meaning. This is particularly true among semioticians. For example, Brier (1995) considers: 'Information is the interpretation of signs by living, feeling, self-organizing, biological and social systems'. This is in direct contrast to the thesis presented in this book, which posits that no science of information can emerge until it is understood that information does not need to be interpreted in order for it to exist.

Meaning is the product of 'information work'. It is a two-step process, although in more primitive information-processing systems the two steps are more or less merged into one. First, the recipient selects a specific piece of information from an environment filled with information. Second, the selected piece of information is analysed so as to extract ***meaning***. Information becomes 'meaningful' when the recipient of information is able to link it successfully to its pre-existing ***internal*** information environment – what, in human terms, is generally described as 'putting it into a context'. However, this process is as true for the metabolic machinery of a bacterial cell interpreting the message encoded in its DNA as it is for a human brain to interpret the message encoded in this sentence.

As implied above, ***information*** must not be confused with ***message***. Information is the raw material which, when information-processed, may yield a message. A message acquires ***meaning*** when it becomes coupled to the internal information environment of the recipient. ***Ascertaining meaning*** is the ***process*** by which a recipient analyses incoming information and discriminates between meaningful and not meaningful information. It represents the first step in the process described above. ***Meaning*** is a ***state*** which is achieved when the incoming information becomes integrated into the information structures already present in the host. This message–context complex (a semantic complex) may, in turn, be further information-processed. An advanced information system such as the human brain may treat the semantic complex as if it were a message even though it has been internally generated. This would allow for the possibility of a second-order message–context complex (the original semantic complex plus its secondary context). By repeating this process, a hierarchy of contexts may be achieved which, if sufficiently elaborate, creates ***understanding***.

This is consistent with the thoughts of C.S. Peirce, who believed that: 'we have to interpret our own thoughts. They are as much signs to us as are the words of other people and the things of the universe'. As O'Connell (1991) states in reviewing Peirce: 'Thinking is talking by myself'. In the process, the human brain draws in more and more of its stored information. If successful, it is able to reorganize existing information into new patterns. This is how the recursive information processing of new (meaningful) complex of information may, at some point, result in creating a new and novel knowledge structure.

It is intuitively evident that merely *decoding* a message requires no intelligence. For example, a radio decodes the incoming electromagnetic radiation and transduces it into sound waves comprehensible to a human listener. We would not ascribe the property of intelligence to the relatively simple electronic circuits of the radio which carry out this decoding. Decoding a message is not the same as ascribing meaning to it. For a message to become meaningful, the information contained by the message must be matched up to the internal information environment of receiver. As with decoding, this can also be a purely mechanical process requiring no intelligence, for example when considering the growth of a crystal or, for that matter, the radio itself, tuned to receive a particular wavelength and ignoring all the others. For the radio, that particular carrier frequency has meaning (the information of that frequency and the internal information environment provided by the radio tuner interact constructively). Thus, contrary to popular belief, it is not necessary for a system to possess intelligence in order for it to either decode a message or ascribe meaning to it.

Meaning, like intelligence, must be understood to represent a spectrum of phenomena. At one extreme is the purely mechanical process exhibited by a growing crystal selecting the appropriate atoms or molecules out of a mixture of substances.

Slightly higher up the scale is an enzyme molecule which discriminates from among hundreds (if not thousands) of different kinds of molecules to enfold a specific substrate molecule. Parenthetically, not only does such an enzyme select a specific set of substrate molecules, it is itself subjected to a variety of control mechanisms which activate or inhibit its tasks. In other words, it is itself being information-processed. Such externally imposed information processing – necessary for the coordination of cell metabolism – results in structural changes of the enzyme molecule. This means that the internal information environment of the protein can be altered: how, or whether, it enfolds a given substrate molecule may be changed. Should an inhibitory message change the resonance structure of the enzyme molecule to the extent that the enzyme has become totally inhibited, it will no longer select the substrate molecules from among the others in its surroundings. The internal information environment of the enzyme can no longer supply a context for the substrate. The substrate molecules will have lost all 'meaning' for the enzyme.

At the other extreme of complexity one might contemplate a philosopher pondering the meaning of meaning.

Thus, ascribing meaning to a message involves a ***spectrum*** of phenomena. At the lower end of this spectrum, the process of ascribing meaning to an incoming message or other form of information does *not* require intelligence. Near the top end, the internal information environment presented by the brain of the philosopher is so rich that it allows the incoming information to be compared with many different aspects of preexisting and organized information which, in turn, allows meaning to be extracted at many different levels. This multiplicity of 'semantic' interactions leads to new knowledge structures and ***insight***.

Defining Meaning

It now becomes possible to attempt a technical definition of ***meaning***:

> Meaning is a state which defines the relationship between two categories of information: external information, which is presented to a recipient system, and pre-existing information stored by the recipient system, which can interact with the incoming information.

The first, if it interacts with the recipient, is usually referred to as a ***message***, and the second is the ***context***. In the present work, this use of the term 'context' has been replaced by the term '***internal information environment***'. The latter, although more cumbersome, is more descriptive. It is the interaction between these two forms of information which determines the extent of the meaning that the message has for the recipient. The greater the interaction, the more bonds are created between the new information and the various information structures existing within the internal information environment of the recipient. The greater this interaction, the more meaningful becomes the information carried by the message.

In one of the simplest of cases, the interaction between an enzyme and its substrate, the interaction is a purely chemical one resulting in the formation of chemical bonds between the two molecules. On the other hand, in the example of the cleaner who finds a piece of a jigsaw puzzle the visual input of information into the human brain results in what might be called ***semantic bonds*** between the mental image of the puzzle piece and the brain's store of information. In Johnny's case, the jigsaw puzzle piece not only created such a set of semantic bonds, but the resulting semantic complex became an internally generated message moving from one gestalt to another.

The creation of enzyme–substrate bonds and of semantic bonds is part of a general process called 'metabolism'. It matters not whether such metabolism is cellular or semantic.

The Message and the Sender

So far, we have focused entirely on the recipient. What about the message itself? What about the sender?

An examination of the evolutionary origin of messages and codes demonstrates that messages and codes existed first, and that meaning came later. Our universe is filled with messages – sunlight for plant shoots to grow upward, gravity for plant roots to grow downward, of pollen grains blown out to sea, radio waves traversing the reader's environment, books on a library shelf – all these are messages that may exist whether there is a recipient or not. Sunlight and gravity existed long before there were plants; the pollen grains will serve no reproductive purpose for which they were designed; the radio waves and their human messages will go unheeded unless the observer turns on the radio – and then tunes in to only one radio transmission at a time; and many a book has lain mouldering on a library shelf unread. It is not until a recipient of these various types of message absorbs the information and lets it interact with its own internal information environment that the message, whatever its form, becomes meaningful.

It harks back to the old question: if a tree falls in the middle of a wood, and no one hears it, has it produced sound? The answer according to the view presented here is this: yes, it has produced sound – the air molecules have been set vibrating, creating what we consider to be the basis of a sound. However, the sound has not become meaningful to any listener because there was no recipient to receive this message from the environment.

No meaning can be ascertained if the environmental information is not coupled to its potential context – real or imagined. Which leads us to the second question. What about the sender of a message?

Here we must distinguish between two categories: a sender who deliberately sends a message with a view to changing the behaviour or status of the recipient, and a sender who sends out a message with no audience in mind, or an entirely different one from the one for whom the message was intended. In the second case we are not dealing with a sender of information, but rather with an emitter or transmitter of information.

This contrasts with the dynamics of a sender who sends a message to a recipient whose behaviour the sender intends to influence. 'Intention' here is a term used rather loosely in that it would apply not only to a human sender, but equally to an ant trying to influence the behaviour of its nest mate, or a gland releasing a hormone. In each case, the sender tailors the message so as to be meaningful to the intended recipients. Thus, questions such as 'what is the meaning of a message?', or 'what did the sender mean when he sent this or that message?', are meaningless if they focus solely on the message itself, or on its sender alone. The meaning of a message, or its intention, cannot be measured without taking into account the internal information environment of the intended recipient.

For example, a business placing a sign over its door which proclaims

PECTOPAH

would not be aiming at customers unfamiliar with the Cyrillic alphabet. For the average Westerner, the sign would have about as little meaning as the piece of jigsaw puzzle did for the cleaner; however, for a hungry Muscovite the sign[1] would not only have a lot of meaning – if she were sufficiently affluent to afford the restaurant and sufficiently hungry – the message conveyed by the sign would be highly *significant.*

Any effort to analyse the ***meaning*** of a message or the intention of its sender becomes irrelevant if it does not focus on the context provided by the recipient, potential or otherwise. The context is, of course, provided by the internal information environment of the recipient, and it is to that internal information environment that the sender tailors the message. Thus, both the meaning of a message and its significance must be ascertained in terms of the intended recipient.

Some Further Thoughts About *Consciousness* and Why it is Possible to Explain it in Reductionist Terms

There was a time when the phenomenon of 'life' was so mysterious – was so beyond our comprehension – that scientists believed that life involved a force very different from the forces of nature encountered in physical systems. This life force was given the name *élan vital* – the vital force – and the theory became known as ***vitalism.***

In 1828, the German chemist Friedrich Wöhler synthesized urea. This meant that an organic chemical, produced by the human body, could be synthesized in the laboratory. Wöhler also helped to establish the field of human metabolism. These were the first of many blows against the theory of vitalism. Still, the idea hung on well into the 20th century.

Much time has been spent in the present work on reviewing aspects of the metabolism of cells. The purpose of such a review was not only to consider the workings of an information-processing system more sophisticated than that of a computer, but also to provide a glimpse into the enormous strides the biological sciences have made in unravelling the

1 which transliterates into the Latin alphabet as 'RESTORAN'

complexities of biosystems. Nowadays, no one considers vitalism to be a cogent explanation of the phenomenon of life, although we must all be mindful of the admonition of Erwin Schrödinger (1944, p. 76): '...the construction [of living matter] is different from anything we have yet tested in the physical laboratory'. This does not mean that biosystems are not subject to the laws of physics, but rather that such laws will involve principles not yet known to us.

Today, to many, the phenomenon of ***consciousness*** appears as mysterious as did the phenomenon of ***life*** two centuries ago. David Chalmers, of the Department of Philosophy at the University of California at Santa Cruz, posits that '...conscious experience be considered a fundamental feature, irreducible to anything more basic' (Chalmers, 1995). He proposes that a future theory of everything will have two components: physical laws, covering physical systems from the infinitesimal to the cosmological, and ***psycho-physical*** laws, which relate to conscious experience. Chalmer makes the analogy to the discovery during the 19th century of electromagnetic phenomena, which could not be explained in terms of previously known principles. As a consequence, electromagnetism, as a fundamental entity, was added to the known forces of nature. Chalmers would ask that: 'Similar reasoning should apply to consciousness'.

The astrophysicist David Layzer, in his thoughtful book *Cosmogenesis*, takes a similar position: '...consciousness... is a brute fact that could not have been predicted... it is emergent in a metaphysical, non-scientific sense' (Layzer, 1990, p. 265).

In the light of the present state of our ignorance, these views are not unreasonable. However, they will prove as erroneous as did vitalism. In fact, we can already see the outline for an explanation of the phenomenon of consciousness.

We begin with the observation of Susan Greenfield on the correlation between neurological structures forming a ***gestalt*** and the thoughts, ideas and perceptions which make up our inner self. Establishing such correlations is a first step. It does not address itself, however, to what Chalmers calls 'the hard problem': nobody knows why '...physical processes are accompanied by conscious experience...' or why '... subjective experience seems to emerge from a physical process'.

There are two parts to attempting to answer this 'hard problem' successfully. The first part invokes an evolutionary perspective. The analyses put forward by Chalmers and many other philosophers tend to be extremely anthropocentric. Much better is the approach of Professor Sloman, as discussed in Chapter 10: 'Instead of asking which animals are conscious and which are not, we might for instance, talk about 75 different kinds of consciousness...', and he goes on to cite some different ones. An evolutionary perspective immediately conveys the idea that as information-processing systems became more complex – as intelligence became more sophisticated – so did consciousness. Consciousness, like intelligence, and like meaning, involves an ***evolutionary spectrum*** of phenomena. To understand consciousness, it is not a good idea to begin by focusing on the incredibly complex upper end of the spectrum.

Instead, one begins with an assumption that simpler information-processing systems such as thermostats have no consciousness. One might extend this assumption to include more complex information-processing systems: biosystems such as single cells, sponges and plants – systems which possess no nervous systems. This assumption can be validated by the following test: can the individual which responds to messages from the environment learn? If not, we may say that there exists no consciousness. (In light of the transgenerational learning which takes place among all biosystems, we might, at some later date, conclude that some degree of consciousness may also be possible as a species property.)

When we come to organisms possessing a nervous system, a new dimension is introduced. The presence of a nervous system allows individuals to learn. Such a system responds not only to incoming information (the stimulus), but is able to correlate such a stimulus with other environmental information inputs. Further up the scale – and this is as true for the ***evolution*** of the nervous system as it is for an advanced system during its ***development*** in the embryo – there is developed an internal information environment sufficiently sophisticated to allow information to be metabolized. It may turn out that the presence of a semantic metabolism is not a sufficient condition for achieving consciousness, but it will certainly prove to be a necessary one.

This business of semantic metabolism leads us into the second part of explaining consciousness – the 'hard problem'. For this we need to recall several concepts discussed in the previous chapters: collective intelligence, synergy, information transduction, virtual reality and information as a basic property of the universe.

Information, as a basic property of the universe, can exist in its own right, independent of any vehicle or carrier which might be transporting or storing it, and it does not need to have meaning in order for it to exist. In Chapter 2 we took the ultimate step in subscribing to this view. We considered the possibility that information, like energy, might exist in a particulate form. Such particles, infons, may or may not be proven to exist at some point in the future; however, there already exist 'virtual particles', such as the holes left behind by electrons moving along a crystalline structure. The holes move but, like the sign at Times Square, there is no change in the physical structure of the system. At the Times Building, the light bulbs do not move; only the information travels. We shall return to this example shortly.

To understand how a physical structure is able to generate information requires us to understand that energy and information are readily interconvertible. Thus, the energy expended by nerve cells (and nerve tissue is among the most energy-consuming of all tissues) becomes, in part, converted to pure information. The neuron has evolved a special system for carrying out just such a transduction. By expending large amounts of metabolic energy, individual nerves are able to conduct impulses along the membranes of their elongated cell body and transmit information across synapses by means of neurotransmitters. This part of the process is well recognized.

What is not as clearly understood is how the totality of this separate information activity becomes transduced into a very different form of information. The brain is an organ which has evolved to carry out such specialized information transduction. As was discussed earlier, just as an electricity network lights up a home by means of a series of ***energy*** transductions,[2] so does the brain represent a system capable of transducing information into its various forms.

To remind ourselves just how efficient the human brain is at this task, let the reader read this paragraph out loud to see how fast the information represented by patterns of dye molecules on this page can become transduced into patterns of sound waves carrying the same meaning.

Returning to the difference between individual nerve impulses and the collective of neurons working in unison: one would not expect a colony of army ants to be the terror of the forest by looking at an individual (rather stupid) ant. This illustrates a general rule: one cannot predict the properties of an ant colony by studying an individual ant; one cannot understand the behaviour of a computer by studying a single chip; and one cannot comprehend the activities of the brain by examining the physiology of a single neuron. Only by invoking the concept of collective intelligence and its product, ***synergy***,[3] do these phenomena become explicable.

Susan Greenfield's ***gestalts*** represent much more than a collection of nerve cells. They are a sophisticated system creating a semantic space by

2 **chemical energy** from burning fuel, to ***mechanical energy*** of the turbines, to ***electrical energy*** via the generators, to ***light*** via the light bulb

3 The whole is more than the sum of its parts.

flashing signals back and forth across the network. The individual signals, millions (or more likely billions) of them, like a collective intelligence, interconnect, fuse with each other, inhibit each other, stimulate each other and form resonating patterns. And out of all this information activity comes something which no longer has mere ***physical*** reality: it has become ***virtual*** reality.

The brain is an organ which has evolved to specialize in creating virtual realities. This is self-evident when we dream. There are the soothsayers who believe that dreams are all part of reality, and they are right in that dreams are a real phenomenon. However, the content of a dream is not reality but a confused manufacture of the brain. Even when we are caught up in a dream so real to us as to lead us to think we are not dreaming, we are still in the grips of the excited brain, which is discharging into the semantic space an incredible amount of detailed and sophisticated information. To quote J. Allan Hobson, Director of the Laboratory of Neurophysiology at the Massachusetts Mental Health Centre: '...each dream is created by the activation of 100 billion neurons in our brain, each connected to at least 10,000 others, and all chatter away at rates up to 100 messages per second...' (Hobson, 1994, pp. 215–16).

If dreams reflect what appears to be a great deal of chaotic information processing, the fact remains that many dreams take up recurrent themes – some clearly of significance to the dreamer. Hobson interprets this phenomenon in terms of complexity theory (Hobson, 1994, p. 216): 'All complex systems – and the brain–mind is certainly a complex system – are characterized by a constant, dynamic interplay between chaos (unpredictability) and self-organization (orderliness)'.

Some, probably most, perhaps all, dreams have some input of consciousness to help organize the flood of information. Here we are referring to two earlier themes. The first is the theme, stressed repeatedly, that information can always be imposed on existing information, thereby creating a more organized form of information. The second, as discussed in the previous chapter, is the idea put forth by Kosslyn that consciousness serves as a kind of parity check. It is this parity check which keeps the restless brain from going mad.

Let us recount the electric news bulletins on Times Square. How can the light bulbs, which are stationary, produce letters that move? All each individual light bulb does is go on and off. It is a simple binary system and no amount of study of individual light bulbs will explain how the letters move. But by coordinating the ***system*** of light bulbs, a moving tickertape of information is generated. The light bulbs generate the information visible to the throngs on Times Square, but that information is a separate entity from the bulbs themselves.

If the lights themselves are not actually moving, then, to the observer, the movement represents a sort of virtual reality. Some might argue that the movement is not a reality at all, merely an illusion. However, if that is an illusion, it must be a very powerful one. Not only have millions of people seen it who can attest to the letters moving across the building like a tickertape, but one can take any individual from any part of the world who, upon looking at the sign, will describe the same phenomenon. As long as the Times Building remains standing and continues to display its tickertape sign, we can rest assured that people seeing the display in an hour, a week or next year will also attest to the fact that the letters move.

The movement is real. It is an emergent information property of the system.

To put it more simply, the movement represents an information output of the system.

This example also illustrates a system engaged in information transduction. The operator inside the building types the information into the electrical machinery, which, through a series of manipulations, coordinates the on/off switches of the light bulbs outside. The nature of the information in front of the operator (printed text) is very different from the information disseminated by the sign, although the meaning of both is the same (approximately). Although the transducing machinery of the human brain is much more complex, in principle the information transduction taking place in the Times Building is similar to the reader reading this printed text out loud.

Another example of what could be trivialized by calling it an illusion, but which represents another form of virtual reality, is illustrated by holography. When we view a hologram, we have the impression of viewing an object in three dimensions. In reality, we are doing nothing of the sort. What we are actually looking at is a flat plate of a photographic emulsion, just like a photograph. Of course, the plate is being illuminated with coherent, monochromatic light (the same as the reference laser beam which took the picture in the first place) and it is this which gives us the illusion of three dimensions. Again, what appears to us as physical reality is really a virtual reality.

Holography is still pretty much in its infancy. In the not-too-distant future it will be possible to have a computer generate a hologram on its monitor in the same way that it is now possible to generate a picture. In either case, one is dealing with forms of virtual reality. In present-day computers the picture projected on the screen has no physical counterpart inside the computer. No matter how hard we look for a picture inside the computer, none is to be found. That will be even more dramatic with a hologram, which would cause us to look for the object itself, which our eyes tell us is there. All we would find (if we knew enough) would be ***patterns*** of transistorized on/off switches representing the information required to project such a hologram.

Both the Times Square sign and the hologram consist of matter and energy. Without them, the information could not be created. However, what emerges is an information output: a coordinated movement of letters across the sign, or the emergence of three-dimensional objects in the hologram. Both this movement and the three-dimensional images are ***real*** phenomena. However, these are phenomena whose reality does not consist of matter and energy; rather, they represent phenomena whose reality consists of pure information. Only by acknowledging that information is a basic property of the universe in the way that matter and energy are does the ***emergence*** of a reality consisting of pure information (that is, a virtual reality which is not based on matter and energy) begin to make sense.

Thus it is with the brain. Except, of course, we are not dealing with a simple tickertape of news bulletins flashed on by patterns of binary electric light bulbs, nor with a more complex hologram, but with an incredibly complex analogue information processor consisting of a hundred thousand million nerve cells. The brain transduces the energy expended by neural activity into a multidimensional semantic space, a space which is based ***physically*** on the separate and independent neurons congregating into a network around an epicentre, but whose ***reality*** (like the moving letters, or the hologram) is an 'information reality'. In turn, the pulsating epicentres become coordinated across the brain to create a higher level of semantic fields. Billions of neurons firing in synchrony, resonating across the brain, result in an enormous output of pure information. It is this form of pure information which we call ***thought***.

Patterns of nerve impulses, which mimic previous information inputs (whether originating externally or internally), create memory. By associating a memory with other memories, plus all new internally generated thoughts and ideas, a kaleidoscope of impressions is created. These impressions are then filtered out, sometimes instantly, some later, probably most during sleep. Just as, in the infant, the mastery of neural patterns mapping light, shape, colour, movement, shade, distance etc. finally becomes organized into mother, pillow, crib, window etc., so does our internal information environment create semantic fields which sort out ideas.

The complexity of our internal information environment reflects the evolution of our human brain as well as our individual and culturally collective experience. Consciousness, as a part of our human internal information environment, has become so subtle and so complex that we become pessimistic of ever unravelling its mysteries. However, as the above and the preceding chapters should make clear, a general theory of information would allow us to understand consciousness and other mental phenomena in terms of the physiology and the physical infrastructure which make up the brain. Consciousness can be reduced to first principles if one considers information in its various manifestations to be an intuitive first principle.

· 13 ·

How Neurons Create Thought

The Brain as a Series of Resonating Circuits

The 'hard problem', however, remains. How do billions of neurons firing in synchrony create consciousness? How does a physical entity, the brain, create an entity as abstract as the mind? How do neurons produce thought?

Information is encoded in a ***book*** as patterns of dye molecules on a printed page. In a ***computer***, the information is encoded as patterns of on/off switches. In the ***brain***, the information is encoded as patterns of neural networks. This is now generally accepted. What is proposed here is that these neural networks are composed of an uncountable number of circuits which resonate. Each such resonating circuit would have the capacity to generate an output wave whose characteristics would be determined by the physical properties of the neural loop. For each semantic unit of information, stored or processed, there would exist in the brain a unique neural loop, or set of loops, emitting a series of pulses or waves whose properties would be unique for that specific semantic unit. Let us explore this concept in greater detail.[1]

---

1 Much of what follows includes fairly standard neurobiology. So as not to lose the thread of the argument, no references are cited. Most of what is covered may be found in the following books: Greenfield (1995); Shepherd (1990); Taylor and Mannion (1992); Wolpaw *et al.* (1991); and papers by Érdi (1996) and Kraus, *et al.* (1996). In addition, some of the work cited here, including Edelman's neural Darwinism hypothesis, has been reviewed earlier by the author (Stonier, 1992).

After this manuscript was written, the book entitled *How Brains Think*, by William H. Calvin, came to the author's attention. Although the conclusions are somewhat different, much of what is covered here is very similar. The reader is urged to read this particular work for another, but closely related, view.

Resonance is a concept which has been alluded to throughout this work (in particular in Chapter 11). It is a phenomenon which occurs at many different levels of organization, from the subatomic to global society.

The acoustic space provided by musical stringed instruments is a well-known example of resonance. This space not only allows for amplification, it also gives each instrument a distinctive character. Similarly, the voice box, bones and sinuses in our heads act to amplify the human voice and give it certain individual resonant qualities. It allows an opera star to shatter a wine glass by loudly singing the note which resonates with the glass.

Technically speaking, a physical system is said to exhibit resonance if a relatively weak driving force, administered at a frequency close to the resonance frequency is able to excite the system into activity for a time much longer than the expected period for internal motion in that state. Resonance phenomena occur when an input of energy is modified by a feedback system, either positive or negative, setting up a rhythm of activity. Such a system acts as an oscillator. Dynamic oscillating systems tend to be more stable than static ones: oscillating systems usually delimit a process, tending to shut off outside interference.

Our universe is filled with oscillators setting into motion cycles which may involve the tiniest fraction of a second, as in atomic and electromagnetic systems, or they may involve daily, monthly or yearly cycles, as in advanced biosystems. To take as an example a physical system, intermediate between these extremes, consider a night-light with a photocell which automatically shuts it off the moment the room is lit up. In a dark room, placing a mirror in front of the night-light causes the light to start flickering: the mirror creates an oscillation alternating between light and dark.

What is happening is that the light reflected by the mirror onto the photocell causes the photocell to shut off the electricity to the light bulb. However, as soon as the light is off the ensuing darkness causes the photocell to stop interfering with the current that activates the night light. The light goes on again. The light reflected in the mirror activates the photocell once again – repeating the cycle. The frequency of the resonating system is determined by the delay in the response time of both the photocell and the light bulb.

Electronic circuits, particularly those designed to send out radio waves, work on a similar principle. What is proposed here is that the bulk of the incredibly large number of neural circuits in the brain are closed-loop resonating circuits. When a pulse of energy is applied to such a neural circuit it oscillates. As a consequence of such oscillations, it is able to produce a sequence of pulses and waves which cycle across the brain like ripples produced when a pebble is dropped into a pond. Just as an input of energy applied to the resonating system of a musical instrument produces ***sound waves***, and just as an input of energy applied to a resonating electronic circuit attached to an antenna produces ***radio waves***, so does an input of energy applied to a resonating neural circuit produce something which we may call ***semantic waves***.

Both individual neurons and individual neural loops may become incorporated into any number of larger loop systems. This means that an individual neuron may participate as a member of tens of thousands, perhaps even millions, of such circuits. The number of resonating systems possible in the brain is so large as to be almost inconceivable. Neurons of the frontal cortex are connected, on average, to 6000 other neurons. Take a 'simple' neural network consisting of only three nerve cells. Such a triad is made up of first, an *input fibre*; second, a *principal* neuron, which acts as an output device; and third, an *interneuron*, a local neuron which is concerned with local processing within the nerve centre. If each of these neurons were connected to 6000 other neurons, then a message entering an input fibre, working through one of the 6000 interneurons, which are, in turn, connected to 6000 other neurons, would be presented with the possibility of traversing 36 million different routes by a simple triad.

If we interpose a second interneuron, itself connected to 6000 other neurons, then we have a four-neuron system which would present 144 thousand million possibilities, while a relatively simple five-neuron system would allow for 864 000 000 000 000 possible trajectories. Now consider that the neural network of the human cortex is estimated to consist of the order of ten thousand million (10^{10}) neurons and sixty million million (60×10^{12}) synapses, and we have a case of the so-called ***combinatorial explosion***. To calculate the number of possible closed-loop circuits in the human cortex would yield numbers so big as to dwarf the very large numbers we associate with describing the distances between stars and the number of stars and galaxies of the universe.

Further complexities arise from the fact that individual neurons may exist in several different states: they may exhibit a stable resting potential; they may be involved in small oscillations, sometimes behaving like a beating heart; or they may be excited into bursts of high action potentials. Depending on the input stimulus, neurons are able to fire in several different modes, stimulating, reinforcing, inhibiting or suppressing the action potential of their neighbours. The synapses which transmit messages across neurons involve another system of exquisite complexity: more than two dozen chemicals acting as neurotransmitters have been identified, while the receptor proteins which complex with these transmitters of information are thought to number in the hundreds. This leads to differences in the synaptic interactions, both in time-scales and in terms of different kinetics. Depending on the topology of the system, therefore, the possibility for varying the dynamics of individual circuits admits for another combinatorial explosion.

Given that kind of complexity, it is easy to ascribe to the brain the capacity to have each semantic unit represented by a resonant system of its own. Not only that, it allows for a huge amount of ***redundancy***. A semantic unit may be represented by many circuits scattered throughout the brain. A single word may exist in association with numerous sensory inputs, such as images, sounds or smells, or it may be associated with larger, more extensive circuits, such as sentences, or with still larger constellations of resonant circuits which represent thoughts and memories. At the same time, because each word (spoken or written) is made up of a unique combination of morphemes or letters, the resonating system representing a word is made up of a unique set of subsystems, each of which represents an individual letter or morpheme. A single semantic unit, in its association with other semantic units, could be stored in a million different places inside the brain.

As reviewed earlier (Chapter 10), a High School graduate in the USA is able to recognize, on average, 80 000 word stems. If each word occupies, on average, a million different sites in the brain, then a High School graduate's entire 80 000-word vocabulary would occupy 80 thousand million (80×10^9) loci. In light of the combinatorial explosion of possible neural networks this does not seem like an outlandishly large number to represent one category of semantic units (words).

When we talk of a semantic unit, we are not only considering the semantic space occupied by a single letter or sound, or of a word, or of a sentence, but any type of message originating in the external environment or generated by the brain internally. This includes images and other visual phenomena, smells, tastes, tactile impressions, temperature sensations and other such external inputs. It also includes internally generated items, simple or complex, such as moods, feelings and thoughts.

Small pieces of information require ***small*** resonating systems. Entire knowledge structures and other complex pieces of information require ***large*** resonant systems. Such large resonant systems are made up not only of a single basic loop of neurons, but rather they contain a hierarchy of resonating systems, each simple unit representing some basic semantic ingredient which becomes part of a larger resonating unit. Such large units, in turn, may be part of a still larger subsystem, ultimately aggregating into the very large integrated system which comprises the knowledge structure. Somewhere, towards the upper end of this spectrum, Greenfield's gestalts must play a role.

To provide an analogy, but using a much simpler system: a radio wave entering a radio receiver is actually a mixture of two sets of waves: a basic electromagnetic carrier wave, and a sound wave. These two combine into a complex wave which is decoded by the receiver. In like manner, the brain is able to mix and decode complex 'semantic waves'.

The Symphony Orchestra Model

Another, perhaps more poetic, model is to compare the brain to a symphony orchestra. Let us begin with a single note produced by a stringed instrument. The shape of the sound chamber (of a violin, cello, guitar etc.) is designed to resonate at certain frequencies. The same instrument can produce hundreds of different sounds: notes, half-notes and all kinds of combinations of these individual notes. Each note, each chord, consists of a resonating system. We now add woodwind, brass, percussion and more up-to-date electronic sound synthesizers. The possibilities allow not only for all the classical music that has been written so far, but all that could be written in the future (providing that there were no further development of musical instruments).

Each musical instrument relies on the same principle: sound is produced and amplified by applying a certain amount of energy to the system (e.g. moving a bow across a string or blowing into a tube) and causing the emerging sound to be amplified by providing a resonance chamber. Different instruments have different kinds of resonating chamber which account for the quality of sound produced. It is these qualities (e.g. strings versus brass) which our ears distinguish easily. In like manner, different neural networks must produce different types of 'sound', that is, ***semantic waves***, whose properties are easily distinguishable by other neural systems which are attuned to these properties.

The capacity of a symphony orchestra to produce a very large array of sounds rests on the capacity of each individual instrument to resonate in its own specific manner. This fact may serve as a model for the production of thoughts in the brain. Like an orchestra playing a large, complex piece of music, a complex thought is based on a combination of semantic waves, each represented by its own resonating circuit. Like the sounds of an orchestra, the resonances produced by the neural circuits combine to form hierarchical systems. By utilizing the orchestra paradigm one may conceive of the process of thinking as involving a symphony of resonating neural networks coordinated into a giant wave of harmonious sound playing complex tunes.

The orchestra model breaks down in at least two respects. First, quantitatively, the brain represents an orchestra composed of at least a million instruments – perhaps a trillion, perhaps much more. Second, a large symphony orchestra normally has a conductor. In contrast, the brain is a self-organizing system which learns the tunes and harmonies as it goes along. However, large orchestras have been known to play without a conductor, and, on a smaller scale, a jazz band, without a leader, will develop pieces of music, having learned them while 'jamming'.

The brain, having a basic anatomy provided by the species' DNA, and beginning to experiment with organizing its neural network while still *in utero*, is able to develop the capacity of an orchestra as described above, but without the equivalent of a conductor. This does not mean that the flows of information are not subjected to prior existing structures and limitations. During the course of evolution, the brain developed numerous organs (e.g. the visual cortex, the olfactory lobes, the hypothalamus) each designed to 'metabolize' different kinds of sensory and bodily information inputs. Mechanisms for coordinating and analysing such information involve providing each unit with genetically programmed algorithms. As with any collective intelligence, the effective functioning of the brain may involve a hierarchy of systems, but, like a colony of ants, requires no 'conductor' to direct complex collective activities.

How the Brain Processes Incoming Information

When we listen to an orchestra, the sound, entering through the ear, is first channelled to the auditory cortex. Here the incoming oscillations (the sound waves interpreted by the vibrations of the hair-like organelles of the inner ear) are analysed, transduced into neural oscillations and categorized. This means that, in its initial phases, the semantic metabolism of an incoming sound involves a search by the auditory cortex for anything which resembles the neural resonances created by the sound.

Existing maps of the auditory cortex indicate that, at one end, the lowest tones are deciphered, while at the other end the highest notes are deciphered. This feature illustrates the principle discussed above: that the brain develops sets of neural resonances not only as a result of experience, but also that it has programmed into it a great deal of information encoded by the individual's own DNA. The human brain, at birth, is not a *tabula rasa* – a clean slate; rather, a baby, at birth, begins its exploration of the environment with a vast array of pre-existing neural network oscillations, and networks with an 'Anlage' – ready to be activated into oscillations given the appropriate stimulus.

More complex sounds coming from the orchestra are analysed further. The string-like sounds are forwarded to centres able to differentiate a violin from a cello; the sounds of woodwinds are sent to centres able to differentiate a flute from a clarinet. To carry out this analysis effectively, the semantic waves (representing incoming information) may be distributed over the entire brain. The process of semantic metabolism requires that an incoming piece of information is matched up with neural circuits whose resonances create similar wave forms – circuits which are a part of the internal information environment provided by the brain. When a semantic wave (representing incoming information) is in tune with some existing resonance it finds a 'home', or, more likely, it finds many such 'homes' – compatible neural networks whose resonances accommodate the incoming oscillations.

It is crucial to understanding that all external information inputs entering the brain must immediately be ***abstracted*** and become transformed into neural oscillations. As with a computer, information inputs must be transduced by the brain in order for it to able to process such information further. The information must be adapted to conform to the brain's protocol for handling information. Inputs set up oscillations in that part of the brain which is specialized to act as the 'first receptor' for such inputs (e.g. the auditory or visual cortexes).

These first receptors probably evolved along with the evolution of the sensor organ (such as the ear or the eye) itself. From an evolutionary perspective, there would be no survival advantage if there evolved a sophisticated ear or eye without, at the same time, evolving a neural structure which is capable of interpreting the signals provided by these sense organs. This means that a great deal of the information transduction taking place in the receptor networks must be carried out by circuits which have been genetically programmed.

After being transduced to a form of information which is compatible with the rest of the brain, the oscillations then become metabolized further by being sent either to other local centres within the receptor organ or to more distant parts of the brain containing stored information of relevance to the input. Such relevant centres contain neural circuits which, because of their appropriate resonance structures, would engage in sympathetic oscillations.

The brain compares and contrasts the characteristics of the initial oscillations created by the incoming information with resonances already stored. At this point the brain acts as a super oscilloscope able to engage in a task of Fourier analysis of monumental proportions. Only the brain's massive parallel processing capability allows it to define the nuances of the complex incoming frequencies.

The final product of all this initial activity is to create a new semantic unit which may either be metabolized further or become stored, or restored, as a memory. The creation of this new semantic unit involves a heuristic process which involves an integration of semantic waves originating in many parts of the brain. This process of integration is carried out in specialized organs or centres of the brain, such as the receptor organs, or, in the case of human speech, in Broca's area. The process works in a heuristic fashion and may result in a direct match-up with previous information or to a mere approximation of it.

A good analogy (and perhaps even a reflection of this heuristic process occurring in the brain) is how we, as a society, name things totally new to us. When we have a new experience, such as a new invention not previously encountered in human history, we create a new word from existing words to achieve a sort of description of the device. In English we tend to rely on Greek and Latin words such as 'automobile' (self-mover), or 'telephone' (voice afar), but in other languages the words may come from the existing vernacular. For example, the French word for railroad is *chemin-de-fer* (track-of-iron), while the German word for television set is *Fernsehapparat* (looking-at-a-distance-device). Such combinations of existing words to describe something new are analogous to the brain assigning neural resonances to new semantic units.

The receptor, or other specialized organ, 'burns' such a new semantic unit into existing neural structures. However, the new circuit has only a limited lifespan. If it is not used, it will decay. Only by repeated use, as we shall discuss shortly, will it achieve some sort of permanence. Unlike a hard-wired electronic circuit, we are dealing with soft bio-tissue, whose synapses (on which all message transmission depends) can be readily modified by altering either the nature or quantity of the two dozen or so neurotransmitters, or the several hundred known receptors at the downstream end of the synapse. Different types of transmitters and receptors are involved in short-term vs. long-term memory. So the brain is able to set up temporary semantic units whose neural circuits may be easily erased, as, for example, during sleep.

Both the brain and the computer are devices which use energy to create, process, store, retrieve and change information from one form into another (information transduction). In the case of the human brain, the oscillating neural circuits represent the physical basis of stored information and, at the same time, allow for extensive and complex information processing by deleting, adding to or modifying existing resonance systems. This differentiates the brain from the classical von Neumann type of computer, where the information is stored as on/off switches and separated from the processing of it.

Meaning Revisited: How the Brain Decides Which Stored Information is Relevant

If the incoming information is entirely new and is transduced into a resonance unknown to the brain, one of two things may happen. If it is totally unfamiliar, it may be ignored. That is, it is excluded from the internal information environment of the brain – the incoming message has no meaning. However, most new incoming information has some connection with something else and may be stored in centres already containing similar, though different, resonances. A new, unfamiliar word may end up in a centre which has a word whose ***sound*** is similar to the new word, or in another which ***looks*** similar in print, or it may end up in a third centre where the ***meaning*** of another word is similar to the new word. Probably it would end up in all three, as well as in others which also create a mnemonic bridge between the new word and the existing internal information environment of the brain. Such existing resonances would then be modified to incorporate the new material.[2]

In an earlier chapter, we introduced the concept of semantic space. It is the confusion between adjacent semantic spaces which leads to the difficulty experienced by patients with deep dyslexia. It is not clear whether such adjacent semantic spaces are actually in close physical proximity to each other in the brain. It may well be. The brain is very good at defining boundaries when activating various networks so that adjacent semantic spaces (represented by resonating neural networks) may be anatomical neighbours in the brain. However, it is also possible that adjacent semantic spaces in the mind are adjacent not because of their anatomical proximity in the brain but because of the similarities in their ***resonance*** structures.

To illustrate this process, children with learning difficulties, when confronted with rapid speech, often cannot discriminate between phonemes such as 'da' versus 'ga', or 'ba' versus 'wa'. According to the theory proposed here, what is happening is that the resonating circuits which

2 This would fit in well with Polya's dictum that a teacher should always present new information in the context of what a child already knows.

represent these phonemes are sufficiently similar to be easily confused by the semantically metabolizing brain.

Like phonons spreading throughout a crystal lattice, so does new information spread throughout the brain. Every incoming oscillation has a chance to be compared to virtually all, or at least most, individual resonances residing in the brain as information waiting to be retrieved. Such stored information is ***self-selecting*** in determining whether or not it is relevant to the incoming information. It takes only a fraction of a second for a stored circuit to recognize the similarity, or lack of similarity, of the resonance properties of the incoming information. This is like a radio receiver, tuned in to incoming electromagnetic radiation: if the resonance is 'incorrect', the message is excluded; if the message is 'in tune', it resonates with the receptor circuit. The receptor circuit becomes activated and the stored information now becomes engaged in a process of retrieval.

Thus we have come back, full cycle, to our concept of 'meaning'. Consider two people conversing. This requires an internal information environment inside the recipient's head which is able to understand the words of the speaker. Without such an internal information environment, the words of the speaker would have no meaning.

But it is not only the listener's internal information environment ***as a whole*** which must be in tune with the speaker. Within that cerebral environment, each morpheme, phoneme, word and thought produced by individual words or strings of words – each of these semantic entities, stored in various parts of the listener's brain – will have to be made to resonate in response to each semantic entity entering the brain.

As an example, imagine that the speaker is reciting Shakespeare's *Hamlet*. Let us concentrate on the representation of the morphemes 'b' and 'ee', the phoneme 'be' and the word 'be', which reside in many parts of the listener's brain. Let us take only the word 'be' and let us consider just one particular locus inside Broca's area of the brain. As the speaker talks, and as the listener attends, the brain is flooded with messages representing the various semantic entities. Our 'be' locus ignores them all ***except*** when the message carries the resonance which activates it. All the other messages have no meaning for our locus.

As Hamlet's 'To be or not to be' soliloquy is spoken, our resonating circuit representing the word 'be' springs into frenzied action, not only because the word appears twice at such short intervals, but to decipher the overall meaning of that statement needs an integration of many parts of the brain. Our circuit will send out messages of its own. And the numerous overlapping semantic fields of which it is a part, will, in turn, respond by creating still larger resonating fields, each of which will also send out further messages.

The brain is constantly being flooded with messages originating both from without and within the brain. The behaviour of our 'be' locus is parallel to the behaviour of an antibody. Both the semantic locus and the action of an antibody involve a process that is ***self-selecting.*** The antibody ignores millions of molecules which it encounters in the bloodstream, but then dramatically engages in a chemical reaction which binds the appropriate antigen into a complex. All the other molecules have no meaning for the antibody – only the specific antigen does. Likewise, the flood of messages traversing the brain have no meaning for our 'be' locus until the right message, consisting of the right wave shape, comes along. Like the antibody, the right piece of information, and only the right piece of information, activates the circuit. For our 'be' locus, only the right piece of information has a ***meaning.***

This principle of self-selection is crucial for understanding the functioning of any form of collective intelligence. We have discussed it earlier in connection with the organs of the body, the growth and development of multicellular organisms and the functioning of ant colonies, and we alluded to it above in connection with the immune system. What it entails is that individual units, comprising the collective intelligence unit, are programmed with a certain set of instructions – algorithms – which are activated by certain stimuli. Such a stimulus, interpreted as a message by the individual unit, may originate either within the collective intelligence system or outside it, as an environmental signal.

The sum of the various organs of the brain constitutes a collective intelligence system. Each organ, in turn, represents a collective intelligence of associated neural networks (there may be several intermediate layers). The point is that the sub-units have been programmed to act of their own accord once they have received the appropriate signal.

Once the 'be' loci have been activated, they will, themselves, send out signals. First, our cortical 'be' loci will send messages back to the auditory cortex so as to allow that portion of the brain to define more precisely the context into which the words and the statement may fit. These return messages help to evaluate and define the incoming information. Second, all the various 'be' loci may combine to create a brain-wide super-resonance. Other 'be' signals may activate more local regions. It is these processes which may be subsumed under the term 'semantic metabolism' – in this case, our word 'be' will become semantically 'digested' to provide either a kind of 'information energy' to the system or, by creating new resonating circuits, become a building block in a new or expanding knowledge structure. This is how the internal information environment of the brain continues to grow.

How Does the Brain Transport Local Oscillations From One Part to Another?

The theory of brain function proposed here requires an explanation as to how it is possible for a particular oscillation representing a sound, image, thought or any other semantic unit to be transposed from one part of the brain to another. That is, assuming that all information is stored as resonating circuits in various parts of the brain, how can the complex properties of a resonating system be transported across the brain to find its counterpart? How is the brain able to match up the incoming information to all relevant bits of information comprising its internal information environment and thereby ascribe meaning to the message?

To answer this question, we begin by referring to the processes which govern a simple radio circuit. Such a circuit consists of a transmitter and a receiver. A radio transmitter contains an electronic circuit which has certain resonance properties. These resonances cause electrons to ascend and descend an antenna with a certain frequency. The alternating current generated causes the antenna to emit electromagnetic radiation at the frequency determined by the resonating circuit of the transmitter. These radio waves are picked up by a receiver, but have a significant impact only if the receiver resonates with the frequency of the transmitter, that is, when the receiver is 'tuned in'. When that happens, the electromagnetic radiation (the radio waves) cause electrons to flow up and down the receiving antenna at the same frequency as the original. (Needless to say, the current of the receiving antenna is only a fraction of the original output of the transmitting antenna; however, if it is 'within range' of the transmitter, a series of amplifying circuits are able to return the signal to a sufficient strength so that it will activate parallel circuits.)

The sequence of information transductions which take place in such a system may be described as follows: the ***information*** represented by an electronic resonating circuit is converted into (1) an alternating current running up and down an ***emitting*** antenna; this produces (2) electromagnetic waves, which result in (3) electrons running up and down in the ***receiver*** antenna, which provides the energy (4) to activate, once more, a resonating system, the receiver – a system which shows properties very similar to that which existed in the original resonating circuit of the transmitter.

This lends itself to considering a first possibility. It may be that a single long axon, going from one part of the brain to another, acts in the way that a radio antenna acts. However, the neuron, unlike the radio antenna, would not send out its signal to the airwaves. Instead, it would interact directly with distant nerve centres. The impulses generated by the transmitting neuron would have to be broken up and timed in such a way that the receiving network would go into sympathetic oscillations. This means that such a transmitting nerve fibre must have the capacity to modulate its output so as to describe the billions of different wave patterns which the theory demands. This is by no means impossible, but there may be a better way.

This second possibility involves a transmission based not on one fibre but a bundle of them. Consider, for example, how 20 fibres could affect a target centre. If each fibre were capable of acting only as an on/off switch (i.e. it were a simple binary system) then 20 fibres would be able to relay 2^{20} possible messages; that is, about a million different messages. Thus, our hypothetical bundle of 20 fibres would be able to transport a binary description of the properties of a semantic wave, utilizing a million bits for doing so.

Of course, the capacity of our hypothetical bundle would be substantially greater than that. Nerve cells are not digital switches – they are delicate analogue devices. Suppose now a simple modification, where each individual fibre has not two but four output states: an 'off' state and three 'on' states, the latter providing pulses of low, medium and high intensity. The possible number of messages that could be transmitted by our hypothetical 20-fibre bundle would now add up to 2^{40}, or about a million million (10^{12}).

Again, we have run into the combinatorial power of the brain's neurological network. A digital transmission by the bundle of nerve fibres described above (and even larger 'mini-columns' of 100 fibres appear to be of widespread occurrence in the cortex) would allow for a detailed description of the frequency and wave shapes produced by an emitting resonating circuit. Such descriptions could involve numerous other details of the properties of the emitting neural network relating to the anatomical and chemical features which characterize the input, feedback and output waves generated by the system, including its location in the brain.

In order for such a system to work, there must be no 'spill over' effect from one fibre to another. Electrical fibres made of copper, for example, not only need to be insulated from their neighbours, but even electrical insulation might not prevent a leakage from a resonating alternating current producing radio waves. Nature, however, has been experimenting with the transmission of information in multicellular organisms for at least several hundred million years. One of the properties of nerve fibres is that they are able to be completely insulated from other parallel fibres. This is true, in part, because the conduction of an impulse along a long fibre does not involve the longitudinal transmission of an electrical impulse. Rather, the message is carried along by the progressive depolarization of the neuron's membrane.

If the evolution of nerve fibres resulted in specialized cells able to avoid a 'spill over' effect, it also created, at the same time, a special organelle: the ***synapse***, a highly sophisticated information gate. Neurons have evolved to become the ideal tool for creating the sophisticated neural networks which underlie the brain's capacity to both store information and transmit it from one part to another.

This emphasizes the difference between information storage and transmission. The former is based on resonating neural circuits, whereas the latter is based on a mechanism we are not yet able to identify with clarity, but if the above mechanisms are operative it is based on moving digitized information along bundles of nerve fibres.

We now arrive at a third possibility: the wave output resulting from the specific resonances of each semantic unit may be added to a ***carrier wave***. Such a carrier wave could be specific to each individual organ or region of the brain, or to constellations of such organs. To make this comprehensible, we return, once more, to our radio transmission analogy.

Each radio station has its own unique carrier wave, and it is by adjusting the resonances of our receiver that we 'tune in' to that particular station. This carrier wave can be modulated by superimposing other waves, such as sound waves, on top of the radio station's carrier wave. The receiver can filter out this carrier wave, leaving intact the sound waves. What comes out of our radio speaker is the sound initially added to the radio waves in the broadcasting studio.

In like manner, the entire brain, or individual organs, or constellations of them, when activated, generate output waves which may turn out to be the equivalent of carrier waves. For example, the ***auditory cortex*** will engage in a 40 Hz oscillation when a human subject is presented with a series of clicks, while the ***olfactory system***, when stimulated, shows bursts of oscillations in the 35–85 Hz range. This contrasts with the 3–12 Hz oscillations in various neural fields of the hippocampus during certain forms of activity, while the thalamus hums along at a slow 0.1–0.2 Hz when its owner is asleep, with bursts of activity in the 7–14 Hz range.

At the moment, it looks as though there might exist a universal carrier wave which diffuses through all parts of the brain, from the ancient brain stem right up to the frontal cortex. Its frequency is based on neural activity, resulting in an oscillation of 40 Hz. This fits in nicely with psychological observations which indicate that the most rapid reaction time to stimuli is around 25 milliseconds, which is, of course, the time it takes to complete one cycle if the system is oscillating at 40 cycles per second.

It is important at this point to recognize that just as there are two different kinds of neural circuits (those which ***store*** information and those which ***transport*** it) so are there two different kinds of wave found in the brain. First, there are those which have been the major concern of this chapter. These are the ***semantic waves*** generated by each specific resonating neural circuit; they represent the coded information of an individual semantic unit. The second type are the well-studied ***brain waves*** discussed in the preceding paragraphs. These seem to be rather nonspecific; they may represent carrier waves concerned with the transport of information from one part of the brain to another. Their existence, however, does not constitute proof that the brain and its various components do, indeed, produce something equivalent to a carrier wave.

Be that as it may, the ubiquitousness and variety of brain oscillations have resulted in a growing feeling among neurobiologists that a detailed study of them will shed light on how the brain actually works, including elucidating the physical basis of consciousness.

Human Speech Revisited

It is interesting to note that human speech, the principal physical output of our thoughts, involves a biologically evolved organ, the larynx, which is itself a resonating system very like a musical instrument. This means that the resonating neural circuit that represents a morpheme can be transferred to a different resonating system made up of membranous folds. The resonances achieved by the vocal cords are monitored by a feedback mechanism to the brain. Such feedback becomes established early in life, when a gurgling baby experiments with producing various sounds. As considered in an earlier chapter, it is at this point that the infant is busy expanding its internal information environment so as to be able to evaluate incoming sounds, particularly those emitted by the human voice.

The idea that the brain induces muscles to perform a highly coordinated series of activities seems not so mysterious to us. We know something about the interface between nerve cells and muscle cells. More importantly, what we observe is a physical system interacting with another physical system. Even though we ought to wonder mightily how all of this information gets coordinated, the problems raised by the brain–body interface are not on a par with the brain–mind problem. The brain–body interaction is no longer of interest to the natural philosophers – it is left to the physiologists to decipher. On the other hand, the brain–mind interaction, as we have seen, continues to engage their serious attention.

Because our historical experience with information has been so limited – covering perhaps only three generations – we have not yet internalized, as part of our cultural wisdom, the idea that information has as much reality as energy. Therefore the idea that a 'material reality', such as the brain, can give rise to something as ephemeral as spoken thoughts remains a total mystery. As emphasized before, only by recognizing that information is as much a fundamental property of our universe as are matter and energy will we be able to solve this conundrum. Only then does the information output of our brain not offend our intuition as we come to understand that the metabolic energy of the neurons and other material structures of the brain can give rise to a form of pure information – a virtual reality – just as the energy of a radio wave amplified by the energy of an added electric circuit, utilizing a television set, is able to transduce the incoming information into a visible 'real' picture. Neither the inside of the television receiver nor the electromagnetic radiation traversing the airwaves contains anything resembling that picture. However, both contain the ***information*** which, when transduced, gives rise to the picture on the screen.

A central theme of the present work is that if we wish to understand a complex mental phenomenon it is best to try to find the origin of such a phenomenon and then trace its evolution. The information outputs of brains of lower animals manifest themselves primarily as contractions of muscles. Muscles become stimulated by neurons which have been activated by messages emitted by the brain. In response to external stimuli, the brain coordinates a variety of complex behavioural responses which cause the animal to engage in exploratory, aggressive, predatory or avoidance behaviour; that is, all the reactions that are part of the organism's behavioural repertoire.

With this as a background, let us return to Chapter 9, on the evolution of human speech. Human speech arose by combining a number of predisposing factors. Among these were the vocalization exhibited by primates and other social mammals. However, to help our understanding of neural aspects of human speech, we should begin by considering the origin of vocalization in vertebrates. For this, we turn to studies of plainfin midshipman fishes. Muscles surrounding the swim bladder of this fish contract to cause the bladder to act like a drum. The muscles can be made to oscillate, thereby creating a low-frequency hum.

Andrew Bass, Professor of Neurobiology and Behavior at Cornell University, has reviewed his own studies and those carried out by other workers (Bass, 1996) on the vocal-acoustic network in the brains of these fishes. Of interest here is that the sonic motor neurons fire at a certain frequency to get the muscles to produce the hum. This frequency is established by two sets of pacemaker cells which lie in close proximity to the motor neurons. The output frequency of the pacemaker determines the frequency of stimulation by the motor neurons, which of course determines the frequency of the muscle contractions. Thus, we see that a semantic wave may be directly responsible for creating a sound wave of identical frequency.

The next step up in the evolutionary process was the development of a vocal apparatus which is effective in air. This brings us to the amphibious vertebrates, such as ***frogs***.

Frogs can make a lot of noise. The chorus of spring peepers and wood frogs is a welcome harbinger of spring. Likewise, the deep call of bullfrogs is well known throughout the summer. The nerves leading from a frog's brain to the muscles of the vocal apparatus are no different from those leading to the powerful hind legs. This means that when a frightened frog jumps into a pond while emitting an alarm call, the interfaces from nerve cells to both the muscle cells of the legs and those of the vocal apparatus have been activated. The output of the brain has caused both body movement and sound – the latter constituting a message understood by other frogs.

Even though a frog's auditory repertoire is very limited, it is a good starting point for devising a model for understanding the output of a brain. The information conveyed by a frog's call tends to fit into one of four categories, which, decoded into human terms, signify 'I am here, all is well', 'I am here, stay off my turf', 'I am here waiting to have sex' and 'Danger, watch out'. This last message gives no information as to the nature of the danger. For that to happen one needs to climb up the evolutionary tree. In Chapter 9 we discussed the variety of calls made by vervet monkeys. The survival value of knowing whether the danger is from a lion or an eagle is self-evident.

In vertebrates, the ability to communicate with a member of the same species resides in that part of the brain called the cerebrum. A frog's cerebrum, although larger than that of a fish, is very small when compared with that of a monkey. In turn, a monkey's cerebrum is small when compared with that of the great apes, and human cerebral power becomes evident when we compare a gorilla's cerebrum, weighing around 430 grams, with our own, averaging around 1350 grams. The cerebrum comprises four-fifths of the human brain; it is the seat of higher mental activities and responsible for our ability to abstract incoming information.

Thus, compared with the information output of a frog's brain, a monkey's brain is very much more sophisticated. In turn, a monkey's capability is small when compared with the information processing and output capacity of the brains of great apes and humans. Late in the evolution of hominids – probably not much more than a few hundred thousand years ago – there occurred an evolutionary discontinuity: the emergence of specialized regions of the brain in the neo-cortex, including Broca's area, which allowed the nerves leading to the vocal apparatus (the mouth and the larynx) to be activated in a highly complex and sophisticated manner. As discussed in Chapter 9, hominid vocalization evolved into human speech.

In principle, the transfer of neural resonances from a frog's brain to its vocal apparatus is no different from that of a human brain expressing its thoughts via the vocal cords. What is different, of course, is the enormous complexity of the human brain's capacity for semantic metabolism, as indicated by the size of the cerebrum. Therefore, instead of being limited to the four kinds of calls of a frog, or the score of known calls of a vervet monkey, the complexity of the human internal information environment allows for a flood of information to be expressed by the human vocal apparatus.

Edelman's Theory of Neural Darwinism Revisited

Patterns of nerve impulses, which mimic previous information inputs (whether originating externally or internally), create memory. By associating a memory with other memories, plus all new internally generated thoughts and ideas, a kaleidoscope of impressions is created. These are then filtered out, some instantly, some later – probably most during sleep. Just as, in the infant, the mastery of neural patterns mapping light, shape, colour, movement, shade, distance etc. finally becomes organized into mother, pillow, crib, window etc., so does our internal information environment create semantic fields which sort out ideas.

Every time a resonating system receives another input, either from an external stimulus or from an internal thought or memory recall, it is strengthened. Strengthening can take place by one of three methods: by increasing the surface area of individual synapses, by increasing the number of synapses, or by recruiting other nerves into the network and setting up a second network to supplement the first. This latter mechanism must take place when a new relationship becomes established between an existing store of information and a second relevant one. Even when such a second network is not created, the first two methods will increase the total area of synaptic contacts within the resonating circuit, thereby facilitating the flow of impulses within the circuit. The stronger the potential for this resonance flow of information, the easier it becomes to activate it, and the less of an input is required to trigger the circuit.

All learned systems, all circuits which were formed as a part of life experience, have a tendency to decay. If not used at all, the decay is more rapid and may lead to the complete disappearance of that circuit as the nerves comprising that circuit are recruited into other more useful ones. The possible exception to this may be those networks most significant to the survival of an individual – circuits which are created by the developing nervous system according to a genetic plan and those learned during an imprinting phase.

A system frequently called upon may modify, and perhaps even displace, other networks whose resonance is close to but not identical to the incoming resonance. This is particularly true as we learn motor skills and as we develop routines and habits. The strengthening of useful circuits not only lowers the threshold at which such a system may be activated but also facilitates the speed at which an input elicits an output, and it reduces the chances for error. In this manner, we become proficient at many tasks, from learning to walk to operating an automobile, from learning how to sing to playing a violin, from learning how to count to mastering calculus, and so forth.

Once Again, Semantic Metabolism and Cellular Metabolism

All neural activity demands, by biological standards, an enormous amount of energy. This energy is derived from cellular metabolism. For sensory and bodily information inputs to be semantically metabolized, that is, for the interpretation of such information and for its storage and retrieval, the brain is wholly dependent on a constant and reliable source of energy.

As any stroke victim knows, interfering with the blood supply to the brain leads to a rapid shutdown of the system. Particularly dramatic is the effect of a blood clot blocking off certain sections of Broca's area, resulting in a condition known as 'expressive aphasia'. The patient appears to have possession of all his faculties but is unable to talk. The patient understands all that is going on around him – even his own condition – and is able to achieve some communication by gesturing. He knows what he wants to say, but the brain, searching for the right words, or even any words, comes up with nothing. Normal oral communication becomes impossible.

If the clot is not too extensive, and if the blood supply has not been seriously impaired for any length of time, the condition may be transient. An extra supply of oxygen will rapidly restore normal functioning. Such a transitory phenomenon provides us with a clue as to what happens during sleep. Like a stroke patient, when we are awakened out of a deep sleep we are often confused. It takes a few moments before we realize where we are (particularly if we awaken in a strange place). If we are asked to respond to questions, it may take a while to come up with an appropriate answer. Sometimes it takes a while to find the right words, particularly if the question involves a foreign language.

How long these deficiencies of semantic metabolism last depends on how fast our body (and brain) becomes aroused. In the case of an emergency, the primitive centres of the brain manage to send signals to the adrenal glands to pump out the appropriate hormones quickly in order to get the advanced centres of the brain to start functioning again.

Deep sleep lowers the brain's cellular metabolism to a point where the level of daytime semantic metabolism is no longer possible. Many neural resonances established during the day are no longer properly maintained and begin to decay. After several nights of this process we will have forgotten trivia: where we parked the car last week in a shopping centre, or what we had for breakfast three days ago. 'Trivial' neural resonances are simply not maintained for any length of time. Like a solar-powered pocket calculator which is not getting enough light, the numbers begin to fade, and finally disappear altogether.

By contrast, neural resonances which are important enough to be retained become relatively permanent by involving one of two processes. Either the neural resonances are used over and over again at fairly frequent intervals, thereby countering the decay, or the resonances were 'burned in' at the time they were learned by being associated with an emotional state which emphasized their importance. In either case, creating an important neural circuit, whether by increasing or strengthening the synapses which comprise part of the resonating system or by creating multiple copies, requires energy. There can be no semantic metabolism without cellular metabolism.

Some Final Words About Information and Meaning

In Chapter 12 we considered the example of the sign 'PECTOPAH'. This sign represents a clear example of the difference between information and meaning. The information contained by the sign is the same, no matter who looks at it, or even if nobody looks at it at all. The meaning, however, is different for the Westerner and for the Muscovite.

It may not be too difficult a task to agree on some yardstick to quantify the ***information*** content of the sign – for example, the number of bits required to define 'PECTOPAH', or perhaps the quantity of energy expended to produce the sign. However, efforts to quantify the ***meaningfulness*** of a message are wholly premature. Not until the internal information environment of a recipient can be described precisely and the interactions between message and context be quantified will it be possible to quantify the semantic value of a message.

At this point in history it is impossible to achieve this for human linguistic systems. However, as the origin and early evolution of human languages becomes fully known, as we learn more about how infants acquire linguistic skills, and as the neuro-circuitry of the human brain becomes sufficiently defined – both in respect of its hard-wiring and its ability to create a linguistic environment – it will become increasingly possible to quantify the interaction between an incoming piece of information and the human internal information environment.

Of great importance will be the contributions made by theoretical biologists, including the attempt to simulate biosystems and subsystems on computers. As Emmeche (1994, p. 87) has pointed out: '...we cannot predict very complex systems: to simulate them directly is the most effective means of determining their behavior'.

It is also hoped that the evolutionary approach promulgated in the present work will make a significant contribution as well. Working one's way up the tree of evolution is always a good experimental approach. As a lesson, one may consider the blossoming of the field of genetic engineering. In the middle of the 20th century there was still some debate as to whether DNA was the carrier of the genetic information. Within two generations, the human genome project was nearing completion. During the early phases, most (although by no means all) significant advances in the science of genetics came from working with microbial systems. This principle, of working on simpler systems first, was still in operation by the end of the century: the organism in which the entire DNA sequence was first uncovered was a bacterium. The second was in a yeast. This latter work, because it involved a eukaryotic organism, was particularly helpful in providing many clues for biochemical geneticists studying the human genome.

In like manner, it is a lot easier to define the internal information environment of an enzyme than that of a sea slug (such as *Aplysia)*. A sea slug is easier to study than an ant. In turn, an ant, whose brain contains only of the order of 100 000 neurons, would present an internal information environment much simpler to study than the human nervous system, which involves almost a million-fold as many. Likewise, to quantify the interaction between the information provided by an enzyme and the information provided by its substrate, and hence its 'meaningful' interaction, would be much simpler to study than what goes on between a grain of sugar and an ant's brain. And, of course, that would be very much simpler than to quantify the semantic interaction between this paragraph and the reader.

The complexity of our internal information environment reflects the evolution of our human brain as well as our individual and culturally collective experience. Consciousness, as a part of our human internal information environment, appears so subtle and so complex that we become pessimistic of ever unravelling its mysteries. However, as the above and the preceding chapters should make clear, a general theory of information would allow us to understand consciousness and other mental phenomena in terms of the physiology and the physical infrastructure which make up the brain. Consciousness can be reduced to first principles, but it requires that information, in its various manifestations, must be considered to be an intuitive first principle.

As to information itself, it has an evolutionary history which dates back almost to the Big Bang, when the various forms of energy began to differentiate out. William Poundstone (1985, pp. 231–2) has defined the universe as 'a recursively defined geometric object' in which 'complexity is self-generating'. To define the various forms of information which have evolved during all this time, to be able to convert one into the other quantitatively, and to derive the laws of information dynamics, we need a Carnot, a Clausius, a Kelvin, a Joule, a Helmholtz, a Gibbs, a Maxwell, a Boltzmann, a Nernst, an Einstein... in short, information science needs to undergo a process of historical development. At the moment, our understanding of these matters is probably as unsophisticated as was our understanding of chemistry before Lavoisier and of biology before Darwin. The present work can merely hope to open the door a crack, hoping that enough light will shine through the door to tantalize others and initiate some serious research in the field.

Appendix A

Calculations Presented in Chapter 2: The Nature of Information

Hartley (1928) postulated that the 'quantity of information' H contained in a message of N signs chosen from an alphabet of S signs could be defined as the logarithm; specifically:

$$H = N \log S \tag{A.1}$$

This equation looked remarkably like that derived by Boltzmann (1896, 1898) in his explorations of statistical mechanics:

$$S = k \log W \tag{A.2}$$

Even though the two equations looked alike, their meaning was very different. This misperception by the traditional information theorists has led to any amount of confusion.

Boltzmann helped to connect changes in entropy to changes in the order/disorder of the system under study. It was this equation which led Schrödinger (1944) to consider the implications of entropy and order for living organisms.

Schrödinger's speculations made it possible to view ***information*** from an entirely different theoretical perspective. Schrödinger modified Boltzmann's equation as follows:

$$S = k \log D \tag{A.3}$$

where S is the entropy of the system, k is Boltzmann's constant and D is the state of *disorder* of the system.

Although Schrödinger did not write it as an equation, it is clear from the text Schrödinger (1944, p. 73) that he considered 'order' to be the inverse of 'disorder', and implicit in his writing is the equation:

$$-S = k \log Or \tag{A.4}$$

where *Or* is the state of ***order*** of the system.

Stonier (1986a,b; 1989; 1990a,b; 1996) proposed that information had to be a basic property of the universe. Specifically, a system may be said to contain information if such a system exhibits ***organization***. That is, just as ***mass*** is a reflection of a system containing matter and ***heat*** is a reflection of a system containing energy, so is ***organization*** the physical expression of a system containing information. By 'organization' is meant the existence of a non-random pattern of particles or energy fields, or more generally, of sub-units comprising any system.

The theory assumed further that there exists a direct linear relationship between the degree of organization and the quantity of structural information. This assumption has proved to be both reasonable and necessary. ***Necessary***, because alternative assumptions lead to unnecessary complexities and intellectual dead ends. ***Reasonable***, because it parallels the assumptions made about the other two parameters which govern our universe: there exists a direct linear relationship between the amount of mass and the quantity of matter; similarly, there exists a direct relationship between the amount of heat and the quantity of energy. On this basis it is not unreasonable to consider that there exists a direct linear relationship between the amount of organization and the quantity of information. Needless to say, the verification of this assumption must await the confrontation of theory with physical reality.

With this caveat in mind, the assumption that there exists a direct linear relationship between the amount of organization *Or* and the quantity of information *I* means that:

$$I = c(Or) \tag{A.5}$$

where *c* represents a constant to be defined.

By substituting in equation (A.4) one may derive the relationship between information and entropy:

$$I = c\,\mathrm{e}^{-S/k} \tag{A.6}$$

The constant *c*, upon further examination, has proven to be equal to the information content of a system when its absolute entropy equals zero. Equation (A.6) was therefore modified as:

$$I = (I_0)\mathrm{e}^{-S/k} \tag{A.7}$$

where I_0 is the information content of the system when $S = 0$.

Equation (A.7) indicates a very different relationship between information and entropy from that understood by the communications engineers. Instead of there being a ***direct*** relationship between information and entropy (as proposed by Shannon and Weaver) or a ***negative*** one (as modified by Brillouin, 1956), equation (A.7) states that information and entropy are ***inversely*** related. To be more precise, entropy is the ***multiplicative*** inverse of information, not, as is generally viewed today, the ***additive*** inverse.

If these postulates are correct, one should be able to calculate a conversion factor. By examining the entropy changes which occur in two different systems – the vaporization of a crystal of ice from 0 K to 373 K and the denaturation of the protein trypsin – it was possible to ascertain the changes in information states and derive a general equation relating entropy units to information (Stonier, 1990). Specifically, the calculations indicated that one entropy unit equals approximately 10^{23} bits/mole, i.e.:

$$1\ \mathrm{J/K} = 10^{23}\ \text{bits (approximately)} \qquad \text{(A.8)}$$

The equation relating ***mass*** to energy shows that very small amounts of matter can create very large amounts of energy. In a like manner, equation (A.8) implies that relatively small amounts of energy can, in theory, be converted into very large amounts of information.

Appendix B
Infons and the Relativity Equations

It is interesting to note that, if the relativity equations are interpreted in a certain way, they contain within them the possibility of the existence of infons. The equation which describes the energy content of a massless particle, i.e. a particle whose rest mass m_0 is zero (as is the case for a photon), is:

$$E = m_0 c^2 / (1 - v^2 / c^2)^{1/2} \tag{B.1}$$

Equation (B.1) implies that if the particle has a rest mass which is zero, i.e. m_0=0, then its energy E must be zero. That is, if m_0=0:

$$E = (0)c^2 / (1 - v^2 / c^2)^{1/2} = 0 \tag{B.2}$$

However, if the particle is travelling at a velocity v equal to the speed of light ($v = c$), then $v^2/c^2 = 1$. If $v^2/c^2 = 1$, then E in equation (B.1) becomes indeterminate, i.e.:

$$E = (0)c^2 / (1 - 1)^{1/2} = 0/0 \tag{B.3}$$

This means that E can have a value, but it cannot be determined by means of equation (B.1). A similar argument is made for 'relativistic momentum', p, which is given by the following equation:

$$p = m_0 \nu / (1 - v^2 / c^2)^{1/2} \tag{B.4}$$

where the energy of the particle is a function of the photon's frequency ν.

Now consider a massless particle moving at a velocity ***other*** than that of light (c). It would possess neither energy nor momentum. Yet, in theory, such a particle could exist. Like a photon, it would possess no rest mass; unlike a photon, however, it would possess no momentum nor would it travel with a velocity of c. Nevertheless, it could possess some other velocity. Such an entity could represent a moving unit consisting of pure information. This would be true even if it were not possible to derive its velocity from the equation above. (Solving for v when both m_0 and p are equal to zero leads to another indeterminate equation with a 0/0 term in it.)

Such 'infons' might not show up in traditional physical experiments, since such particles would possess neither mass nor energy. As such, they would not interact with any detectors that are based on the interaction of an energetic particle with atoms of matter. They could, however, manifest their effect by bringing about changes in the organization of the matrix within which they are operating. Detecting infons would be even more problematic than detecting neutrinos.

Appendix C

How to Evolve Electronic/Mechanical Systems into the Equivalent of a Living, Self-Reproducing Bio-System (a Self-Replicating Automated Factory)

Over the last few decades there has been a trend to automate virtually all record-keeping and all sorts of routine business transactions. This trend has occurred not only in the field of commerce, but also, increasingly, in processing and manufacturing systems, and on farms and in mines. At this point in history, total automation in manufacturing systems is theoretically possible, although economically it would not be cost-effective.

With this as background let us engage in the following thought experiment. Consider, initially, a fully automated automobile factory. Then replace this with a factory which produces, instead, several classes of robots. Finally, have it produce not only different kinds of robots, but all the components for building another factory.

Part 1: Conceptually, Consider a Totally Automated Factory System Producing Automobiles

a. The factory is automated to the extent that it has troubleshooting and self-repair capabilities. Automated diagnostic procedures are standard practice in many garages. Again, there is no theoretical reason why automated diagnostic techniques may not be improved to the point where they equal human capabilities. In extreme cases, where the fault cannot be diagnosed or repaired, the faulty car is simply pushed off the assembly line and scrapped. The same principle that applies to the techniques for diagnosing faults with individual cars can be extended to an assembly line producing cars and to an entire factory.

b. Production involves the creation of interchangeable parts. The interchangeability allows the factory to produce different models. Production also allows for a variety of colours.

c. The factory includes an office which can receive orders (via some form of electronic mail). The orders may be for various models and colours. The office has the electronic (computer) means to transmit this information to 'supervisory' computers on the factory floor, which then alter the production procedure.

d. The office also maintains a parts and raw materials register and orders parts and raw materials as needed.

e. The main factory (producing the automobiles) has a series of satellite factories which produce needed components, such as tyres, sheet metal, glass windows and plastic liners. All satellite factories are also fully automated along the lines indicated for the main factory.

f. The satellites may themselves have satellites. If, for example, a satellite produces engines, it may have one satellite producing engine blocks, another pistons, a third valves etc. Satellite activity may take place under one roof.

g. The satellite factories may become involved with mines, rubber plantations or other resource extraction activities so as to acquire the raw materials necessary for the production of the various components. In some cases, such extractions could also become fully automated. Where this is not possible, the input will have to be supplied by human systems. If the automated factory complex were to be viewed as an organism, then the factors which would be supplied by the human system should be viewed as 'food'. Special components, rather than steel, oil, plastics or other major material substances required by the normal functioning of the factory, should be classed as 'vitamins' or other special growth factors. An organism as complex as a human being, in addition to requiring a large input of energy-producing foods, requires 21 amino acids and numerous vitamins and growth substances for normal functioning. Even lowly bacteria often have a variety of needs for special molecules to be able to grow and multiply. Thus, in an analogous manner, what the automated factory cannot provide for itself should be viewed as the equivalent of special nutrients.

h. All transportation is automated.

i. The whole system, from extraction to car production and sales, could be governed by a master 'company' computer using input/output analysis to coordinate the entire enterprise. However, such a company computer may be quite unnecessary: as with a colony of ants, the intelligence may be ***distributed***. By utilizing fairly simple algorithms, each satellite or primary resource factory goes into production only when ordered to do so by the factory further downstream in the production process. On the other hand, a master 'company' computer could anticipate bottlenecks better and might make a major contribution to the smooth functioning of the system.

Part 2: Instead of Manufacturing Cars, Have Such a Factory System Manufacture Robots

The main factory must be able to create several different kinds of robot. Some of these may be involved in their own manufacture, but others must go out to the satellite factories, mines and wherever else they are required to provide for the production of necessary parts and primary materials.

Part 3: Expand the Factory to Include the Capacity to Produce, in Addition to Robots, Robot-Manufacturing Machinery, and Materials for Creating Another Robot-Manufacturing Factory

a. The robot-producing factory would be expanded to produce a second class of robots. These would not be involved in the running of the factory, or in producing other robots, but would be used for carrying out the numerous tasks involved in the construction of a new main factory and its satellites.

b. To back up this second class of robots, a second satellite factory system would be required. The first, as with the car manufacturer, would be required to furnish the main robot factory with materials and components necessary for the production of ***robots***. The second would provide the materials necessary for the creation of new robot ***factories***, including robot-producing machine tools.

c. For such an effort one would need a hierarchy of coordinating computers which would not only coordinate all production processes (as with cars or robots), but would also duplicate and pass on the blueprints necessary for designing the daughter factory.

Further Thoughts

The above indicates that it is possible to create a system, based only on mechanical systems guided by electronic (or opto-electronic) intelligence, whereby an automated factory could reproduce itself without any kind of human intervention. Although this would be a complex undertaking, if enough resources were committed to the project it would be no more difficult than, let us say, it would have been in 1950 – before Sputnik – to walk on the Moon within twenty years. At the moment there exists no economic imperative to build such systems. However, the situation may change at some point in the future if it becomes desirable to establish mining or manufacturing operations on the Moon, the asteroids or other life-hostile bodies in our Solar System.

Leaving aside such commercial considerations, the main reason for engaging in the above thought experiment at this time is in the context of theoretical biology. In many ways, such a fully automated robot factory, able to reproduce itself, parallels an ant colony. In an ant colony, the specialization of workers and the use of distributed intelligence to allow the whole system to function represent a parallel with individual robots and robot complexes. The factory and the ant nest produce more individual robots or ants and provide a basis for the whole factory or nest to reproduce itself.

There is, however, a profound difference from the ant colony which makes the self-reproducing robot factory more akin to an individual cell than to an ant nest: ants, in taking care of their young, merely feed the larvae – they do not create them (except possibly the eggs laid by the queen). In contrast, the robots on the assembly line actually participate in creating more of themselves. Therefore a better analogy than the robots resembling ants (although that may be the case in some instances) would be to equate them with enzymes inside a cell. The assembly line might be the equivalent of ribosomes and other cell organelles, and the roof, floor, walls, beams etc. might be equivalent to the cell wall and structural proteins, such as collagen.

All life forms (with the possible exception of viruses) achieve their growth as a result of ***intussusception***. Crystals and human manufacturing systems, on the other hand, grow by ***accretion***. The above thought experiment results in a non-biological materials system which grows by intussusception. The automated factory system envisaged above, able to produce other factory systems in its own likeness, fulfils Teilhard de Chardin's criterion of a 'corpuscle'.

The 97 per cent of human DNA that does not code for proteins is currently often referred to as 'junk' DNA, 'genomic parasites' or 'selfish' DNA (Schimenti, 1995). The above description of a non-biological system allows for a better analysis of the information flows. The 3 per cent of DNA which codes for the enzymes and other proteins is comparable to the instructions which produce robots and other structural components of the factory.

Such instructions are only a small part of the operation of the whole system. To properly coordinate the various activities and ensure the production of daughter factories involves higher levels of information storage and processing. The junk DNA therefore must not only contain a great deal of vital information, it is, in evolutionary terms, likely to be of vastly greater organizing significance!

Appendix D

Some Cardinal Differences in the Anatomy and Physiology of Classical Computers and the Human Brain

Computer	Human Brain
1. Digital information processor based on circuits of binary switches	1. Analogue information processor involving a complex nervous system with scores of chemical neurotransmitters and modifiers
2. Information transported as pulses of electrons along conductors, and across semiconductors	2. Information transmitted as pulses of depolarization along membranes and as neurotransmitters across synapses
3. Speed of pulse transmission *ca.* 10^{10} cm/sec	3. Speed of pulse transmission *ca.* 10^{3} cm/sec
4. Relatively simple circuitry but in creasing in complexity	4. Extremely complex circuitry: 10^{11} neurons with up to 10^{15} connections
5. Crystalline structure, extremely stable	5. Bio-tissue vulnerable to damage
6. Can operate under a wide variety of conditions	6. Needs carefully regulated environment to operate
7. Computer system may be shut down indefinitely with no damage	7. Brain requires continuous energy inputs in order to maintain the living system

8. No self-repair. Some self-correction and bypass of faulty areas	8. Tissue capable of significant self-repair. Also extensive capability to transfer function to other circuits
9. Memory based on patterns of binary switches	9. Memory based on patterns of neural connections

Adapted from Stonier (1984, p. 18)

The above table indicates that the computer, in its broadest sense, may be considered as a complex, aperiodic crystal with certain unique electronic properties. Its circuitry is based on semiconductors plus the conductors necessary to complete the system. In principle it is quite different from, and substantially less sophisticated than, the human nervous system. As Margaret Boden (1987) has pointed out: 'AI has shown us that we are much cleverer than we previously thought' – a consideration reinforced by the work of Miller and Gildea (1987), demonstrating that by means of contextual analysis, children, in the years between infancy and late teens, acquire a word-recognition capability of 80 000 words, learning to recognize on average more than 10 new words per day.

Obviously, the human brain has highly complex and effective circuitry and much larger memory storage, which is based on a network of 100 thousand million (10^{11}) neurons, each connected to perhaps a thousand other neurons. In some instances, for example in the case of the Purkinje cells of the cerebellar cortex, each cell receives inputs from as many as 200 000 parallel fibres (Diamond, 1990). Furthermore, each neuron is a sophisticated information device in its own right – much more sophisticated than a transistor, or even a transputer. Think of a 'Connection Machine' computer consisting of 10^{11} transputers and we may approach the capacity of the human brain. In addition to all this biological networking power there remains the possibility that individual molecules, such as proteins or nucleic acids, are also able to store information. It is known that in immune systems individual molecules retain information on the surface of specific antibody cells. The human brain also consists of tissue capable of significant self-repair or, alternatively, compensatory activity to minimize damage to the system.

On the other hand, computers possess advantages in at least three areas: the speed of pulse transmission, the tough and stable nature of the material and (unlike a living system) they can be shut down indefinitely without damage. In addition, the ease of tying computers together – communicating with each other in a way not currently possible for human brains to do so – creates a computer 'collective intelligence'. Already the online mass storage capability for some computers is in the gigabyte range. This range is extendable. Think of every household with a telephone possessing a powerful computer tied into a global network.

Classical computers differ profoundly from the human brain in at least two other respects: (1) the separation of processing and memory functions and (2) a total reliance on logic. However, logic is a form of thinking which appeared only very late in the evolution of intelligence. It also appears late in the development of a child's thinking processes. Most higher animal and human intelligence relies on simple association. Our mental map of the world consists of associating objects, sensations, events etc. into a pattern of memories. These constitute the map which allows us to make sense of the world in which we live. How such a neurological map is created in the first place requires an understanding of current theories of brain development, a matter which has been discussed in a previous work (Stonier, 1992) and will not engage our attention at this time.

Appendix E

Possibilities Considered in Chapter 11 on How to Quantify Meaning

Possibility 1. ***Meaning*** could involve a ratio of *the amount of information carried by a message* to *the amount of information processing carried out by the receiver in response to the message.*

This would be based on a measure of the 'successful' information processing following the receipt of a message. Successful information processing is the opposite of an ***unsuccessful*** random search trying to match up the message with relevant epicentres located in the internal information environment. Such an unsuccessful search would represent a baseline of zero in terms of the amount of meaning a message has for the recipient. On the other hand, the more successful the information processing, the greater the interaction between message and internal information environment. Such an interaction might be treated mathematically in terms of set theory using Venn diagrams to measure the overlap between the information contained by the message and the integrating information provided by the internal information environment – information which would constitute the context.

One might use the example of the jigsaw puzzle referred to earlier, with an effort to quantify how much meaning the puzzle piece held for each of the four people who viewed it. Trying to use Venn diagrams at this point leads to the difficulty of ascribing a value to each of the two subsets – how much information can one ascribe to the jigsaw piece, and how much relevant information exists inside the heads of each of the four characters, representing the quantitative value of the context into which the jigsaw piece fits?

Perhaps meaning might be quantified by measuring the length of the common boundary between the two subsets. Again, this seems hopeless at the moment, because how can one quantify this boundary?

A better bet is the suggestion by Susan Greenfield (1995, p. 95) that the degree of ***consciousness*** '...can be the product of... the power of the epicenter [which] can be defined in terms of the number of neural associations recruited'. It may actually be that consciousness in this case is equivalent to ***meaning***. In fact, this could be one solution to quantifying meaning: how many neurons, or perhaps how many synapses, are activated by an incoming message?

Possibility 2. To try to measure the fate of the incoming information. Thus a 'first-order semantic reaction' involves only a simple interaction between message and recipient – as with inorganic systems. A 'second order' would be achieved when the message–context complex itself acts as a message to be further information-processed into a message–context–second context information unit. A 'third order' would allow the 'second context derived information unit' to be processed into a message–context–second context–third context unit, and so forth, creating new knowledge structures.

Possibility 3. Meaning may be assigned a grade, which would represent the hierarchical nature of the process. A molecule of sugar entering a growing crystal causes information processing to occur primarily as a local response. The sugar molecule, acting as the message, affects the information content of the crystal (change in organization) primarily by changing the bond structure of its neighbours. There may be some transmission of information across the entire crystal in terms of moving electrons, phonons and other organizational properties, but the greatest effect is local. This would be classed as a first-order semantic reaction.

With unicellular biosystems, the information gained from the molecule of sugar is no longer local but is spread throughout the entire cell. The information feedback loops in the cell consist of chemical messengers which react throughout the cell. This would be a second-order semantic reaction. It reflects the difference between the nature of the growth of crystals and cells, i.e. *accretion* vs. *intussusception*.

With multicellular organisms lacking a nervous system, the import of the message can be transmitted throughout the whole organism by means of hormones. Thus a plant receiving a message from the environment that it is early summer (the message consisting of the relative length of the day/night cycles) may change its hormonal status so as to engage in flowering. By involving an entire organism, this would be classed as a third-order semantic reaction.

In multicellular organisms which possess a nervous system but lack a brain, the organismal response is much faster than one which depends on the diffusion of hormones. This, then, would become a fourth-order semantic reaction. Organisms which have brains can remember the source of the message and the circumstances under which such a message might have been received; therefore this would be classed as a fifth-order semantic reaction.

Assigning a ***gradation*** would clarify the difference in the meaning that a message might have for different levels of system complexity. As we move up the evolutionary scale – inorganic, organic, cellular, organismal, nervous and brain systems – the capacity of the system to achieve higher orders of meaning reflects the increase in the system's ability to engage in increasingly complex information processing.

This principle should not be confined to biological evolution, but would apply equally to other evolving systems, such as the cultural evolution of a society or the educational evolution of an individual. For example, the appearance of a comet or the outbreak of an epidemic had very different meanings for medieval society when compared with our own. In like manner, a newspaper would have a different meaning for someone who cannot read compared with its meaning for someone who is literate.

Such a system of grading would have some utility in defining more precisely the spectrum of meaning. Unfortunately, it would give us only a rough outline and would not be helpful in defining the quantitative difference between the meaning that the piece of jigsaw puzzle had for the immigrant labourer and Johnny.

Possibility 4. In physical systems, major increases in entropy are associated with altered physical states, which seem to be associated with breaking certain kinds of bonds or linkages. Might a more fruitful approach be counting the number of bonds, and giving values to the strength of the bonds? Is it possible to obtain similar quantities associated with other forms of information changes in more complex systems?

Possibility 5. Using the concept proposed by Hinton and co-workers, that of 'point attractors' in 'semantic space' (as discussed earlier) as a basis, would it be possible to quantify such interactions?

References

RJ Andrew (1965) The origin of facial expressions, *Sci. Am.*, 213(4):88–94.

DW Anthony (1996) Shards of speech, *The Sciences* (New York Acad. Sci.), 36(1):34–9.

S Artavanis-Tsakonas, K Matsuno and ME Fortini (1995) Notch signaling, *Science*, 268:225–32.

S Aukstakalnis and D Blatner (1992) *Silicon Mirage*, Peachpit Press, Berkeley, California.

M Barinaga (1992) 'African Eve' backers beat a retreat, *Science*, 255(5045):686–7.

M Barinaga (1996) The bittersweet truth about gustducin, *Science*, 272(5269):1742–3.

AH Bass (1996) Shaping brain sexuality, *Am. Sci.*, 84(4):352–63.

C Blakemore (1988) *The Mind Machine*, BBC Books, London.

M Boden (1987) Artificial intelligence: cannibal or missionary?, *AI & Society*, 1(1):17–23.

P Bogh Andersen (1995) Morphodynamic models of communication, in *The Emergence of Codes and Intentions as a Basis of Sign Processes*, informal publication of an interdisciplinary symposium, 26–28 October 1995, Odense University, Denmark.

FF Bolander Jr (1995) Endocrinology, molecule, in *Molecular Biology and Biotechnology* (ed. RA Meyers), VCH, New York, pp. 277–84.

L Boltzmann (1896, 1898) *Lectures on Gas Theory* (Part I and Part II) (transl. SG Brush, 1976) University of California Press, Berkeley, California.

JT Bonner (1980) *The Evolution of Culture in Animals*, Princeton University Press, Princeton, New Jersey.

S Brier (1992) Information and consciousness, *Cybernetics and Human Knowing*, 1(2/3):71–94.

S Brier (1995) Cybersemiotics: a paradigm integrating pragmatic linguistics, second order cybernetics with the triadic semiotics of C.S. Peirce, in *The Emergence of Codes and Intentions as a Basis of Sign Processes*, informal publication of an interdisciplinary symposium, 26–28 October 1995, Odense University, Denmark.

L Brillouin (1956) *Science Information Theory*, Academic Press, New York.

AS Brooks (1988a) Mousterian, in *Encyclopedia of Human Evolution and Prehistory* (eds. I Tattersall, E Delson and J Van Couvering), Garland, New York, pp. 360–3.

AS Brooks (1988b) Paleolithic, in *Encyclopedia of Human Evolution and Prehistory* (eds. I Tattersall, E Delson and J Van Couvering), Garland, New York, pp. 415–19.

SG Brush (1983) *Statistical Physics and the Atomic Theory of Matter*, Princeton University Press, Princeton, New Jersey.

H Bussey (1996) Chain of being, *The Sciences* (New York Acad. Sci.), 36(3):28–33.

AG Cairns-Smith (1986) The four crystal genes, in *Clay Minerals and the Origin of Life* (eds. AG Cairns-Smith and H Hartman), Cambridge University Press, Cambridge, pp. 143–52.

WH Calvin (1996) *How Brains Think*, Basic Books, New York.

J Campbell (1982) *Grammatical Man: Information, Entropy, Language, and Life*, Simon and Schuster, New York.

LL Cavalli-Sforza (1991) Genes, peoples and languages, *Sci. Am.*, 265(5):104–10.

D Chalmers (1995) The puzzle of conscious experience, *Sci. Am.*, 273(6):80–6.

S Chang and TE Bunch (1986) Clays and organic matter in meteorites, in *Clay Minerals and the Origin of Life* (eds. AG Cairns-Smith and H Hartman), Cambridge University Press, Cambridge, pp. 116–29.

C Cherry (1978) *On Human Communication*, 3rd edn, MIT Press, Cambridge, Massachusetts.

ER Cohen (1996) *The Physics Quick Reference Guide*, American Institute of Physics, Woodbury, New York.

J Cohen (1995) Novel center seeks to add spark to origin of life, *Science*, 270:1925–6.

A Combs (1995) *The Radiance of Being*, Floris Books, Edinburgh.

M Conrad (1996) Cross-scale information processing in evolution, development and intelligence, *BioSystems*, 38(2,3): 97–109.

EH Davidson, KJ Peterson and RA Cameron (1995) Origin of bilaterian body plans: evolution of developmental regulatory mechanisms, *Science*, 270:1319–25.

P Davies (1987) *The Cosmic Blueprint*, Heinemann, London.

R Dawkins (1976) Memes and the evolution of culture, *New Scientist*, 72:208–10.

K Devlin (1991) *Logic and Information*, Cambridge University Press, Cambridge.

A Diamond (ed.) (1990) *The Development of Neural Bases of Higher Cognitive Functions*, Ann. New York Acad. Sci. Vol. 608.

W Dowell (1996) Deep Blue Funk, *Time*, 26 February 1996, pp. 60–1.

C de Duve (1995) *Vital Dust*, Basic Books, New York (see also The beginnings of life on Earth, *Am. Sci.*, 83, 423–37 (1995).

C Emmeche (1994) *The Garden in the Machine*, Princeton University Press, Princeton, New Jersey.

P Érdi (1996) The brain as a hermeneutic device, *BioSystems*, 38(2,3):179–89.

R Forsyth and C Naylor (1985) *The Hitch-Hiker's Guide to Artificial Intelligence*, Chapman & Hall/Methuen, London.

Z Galcheva-Gargova, KN Konstantinov, I-H Wu, FG Klier, T Barret and RJ Davis, Binding of zinc finger protein ZPR1 to the epidermal growth factor receptor, *Science*, 272:1797–802.

GG Gallup Jr (1975) Towards an operational definition of self-awareness, in *Socioecology and Psychology of Primates* (ed. RH Tuttle), Mouton, The Hague, pp. 309–41.

M Gardner (1960) *The Annotated Alice*, Bramhall House, Clarkson N. Potter, New York.

RA Gardner and B Gardner (1969) Teaching sign language to a chimpanzee, *Science*, 165:664–72.

L Gatlin (1972) *Information Theory and the Living System*, Columbia University Press, New York.

A Gibbons (1992) Mitochondrial Eve: Wounded but not dead yet, *Science*, 257:873–5.

A Gibbons (1993) Pleistocene population explosion, *Science*, 262:27–8.

WW Gibbs (1994) Software's chronic crisis, *Sci. Am.*, 271(3):86–95.

J Goodall (1971) *In the Shadow of Man*, Houghton Mifflin, Boston, Massachusetts.

SA Greenfield (1995) *Journey to the Center of the Mind*, WH Freeman, New York.

W Greiner and H Stocker (1985) Hot Nuclear Matter, *Sci. Am.*, 252(1):58–66.

DR Griffin (1984a) *Animal Thinking*, Harvard University Press, Cambridge, Massachusetts.

DR Griffin (1984b) Animal Thinking, *Am. Sci.*, 75(5):456–64.

K Haefner (1988) *The Evolution of Information Processing*, Faculty of Mathematics and Informatics, University of Bremen, Germany.

K Haefner (1991) *Evolution of Information-Processing Systems*, Project Evolution of Information Processing, University of Bremen, Germany.

RVL Hartley (1928) Transmission of information, *Bell Syst. Tech. J.*, 7:535.

H Hartman (1986) The clay hypothesis, in *Clay Minerals and the Origin of Life* (eds. AG Cairns-Smith and H Hartman), Cambridge University Press, Cambridge, pp. 10–12.

RL Harvey (1994) *Neural Network Principles*, Prentice-Hall International, Englewood Cliffs, New Jersey.

B Hayes (1994) The magic words are squeamish ossifrage, *Am. Sci.*, 82:312–16.

S Haykin (1994) *Neural Networks*, Macmillan College Publishing Co., New York.

GW Hewes (1973) Primate communication and the gestural origin of language, *Current Anthropol.*, 14(1–2):5–21.

GE Hinton, DC Plaut and T Shallice (1993) Simulating brain damage, *Sci. Am.*, 269(4):76–82.

JA Hobson (1994) *The Chemistry of Conscious States*, Little, Brown and Co., Boston.

C Holden (1996) Random samples: Eavesdropping on ants, *Science*, 272(5264):957, 959.

BK Hölldobler and EO Wilson (1977) Weaver ants, *Sci. Am.*, 237(6):146–54.

BK Hölldobler and EO Wilson (1990) *Ants*, Springer-Verlag, Heidelberg.

J Horgan (1996) Plotting the next move, *Sci. Am.*, 274(5):16.

F Hsu, T Anantharaman, M Campbell and A Nowatzyk (1990) A grandmaster chess machine, *Sci. Am.*, 263(4):18–24.

I Illich and B Sanders (1989) *The Alphabetization of the Popular Mind*, Penguin, London.

C Jensen (1996) *No Downlink* (transl. B. Haveland), Farrar, Straus, Giroux, New York.

HJ Jerison (1986) The perceptual worlds of dolphins, in *Dolphin Cognition and Behavior: A Comparative Approach* (eds. RJ Schusterman, JA Thomas and FG Wood), Lawrence Erlbaum Associates, Hillsdale, New Jersey, pp. 141–66.

A Jóhannesson (1949) *Origin of Language*, H.F. Leiftur, Reykjavik.

D Kaiser (1996) Bacteria also vote, *Science*, 272(5268):1598–9.

S Klein, N Takahata and FJ Ayala (1993) MHC polymorphism and human origins, *Sci. Am.*, 269(6):78–83

T Kohonen (1988) An introduction to neural computing, *Neural Networks*, 1:3–16.

A Kortlandt (1973) Comments following GW Hewes, Primate communication and the gestural origin of language, *Current Anthropol.*, 14(1–2):5–24.

SM Kosslyn and O Koening (1992) *Wet Mind: The New Cognitive Neuroscience*, The Free Press, New York.

N Kraus, TJ McGee, TD Carrell, SG Zecker, TG Nicol and DB Koch (1996) Auditory neurophysiologic responses and discrimination deficits in children with learning problems, *Science*, 273(5277):971–3.

P Krüger (1981) *Wladimir Iwanowitsch Wernatskii*, Biographien hervorragener Naturwissenschaftler, Techniker und Mediziner, vol. 5, BSB B.G. Teubner, Leipzig, Germany.

H Kummer (1982) Social knowledge in free-ranging primates, in *Animal Mind – Human Mind* (ed. DR Griffin), Springer-Verlag, New York, pp. 113–30.

J Lanyi (1995) Bacteriorhodopsin, in *Molecular Biology and Biotechnology* (ed. RA Meyers), VCH, New York, pp. 68–70.

D Layzer (1990), *Cosmogenesis*, Oxford University Press, Oxford.

HS Leff and AF Rex (1990) *Maxwell's Demon: Entropy, Information, Computing*, Princeton University Press, Princeton, New Jersey.

P Lieberman (1988) Voice in the wilderness, *The Sciences*, (New York Acad. Sci.), July/August, 23–9.

F Machlup (1983) Semantic quirks in studies of information, in *The Study of Information* (eds. F Machlup and U Mansfield), John Wiley & Sons, New York, pp. 641–71.

AL Mackay (1986) The crystal abacus, in *Clay Minerals and the Origin of Life* (eds. AG Cairns-Smith and H Hartman), Cambridge University Press, Cambridge, pp. 140–3.

PF MacNeilage, MG Studdert-Kennedy and B Lindblom (1993) Hand signals, *The Sciences*, (New York Acad. Sci.), January/February, 32–7.

EA Marcus and TJ Carew (1990) Ontogenetic analysis of learning in a simple system, in *The Development of Neural Bases of Higher Cognitive Functions* (ed. A Diamond), Ann. New York Acad. Sci. Vol. 608., pp. 128–45.

L Margulis (1970) *Origin of Eukaryotic Cells*, Yale University Press, New Haven, Connecticut.

P Marijuán (1993) Enzyme, automata, and artificial cells, in *Computing with Biological Metaphors* (ed. RC Paton), Chapman & Hall, London.

P Marijuán (1994) Opening Address to the First Conference on Foundations of Information Science, Madrid, 11–15 July 1994 (reprinted in *SIGBIO Newsletter*).

P Marijuán (1996) 'Gloom in the society of enzymes': on the nature of biological information, *BioSystems*, 38(2,3):163–71.

AN Meltzoff (1990) Towards a developmental cognitive science, in *The Development of Neural Bases of Higher Cognitive Functions* (ed. A Diamond), Ann. New York Acad. Sci. Vol. 608., pp. 1–25.

E Menzel (1975) Natural language of young chimpanzees, *New Scientist*, 16 January:127–30.

EW Menzel (1986) How can you tell if an animal is intelligent?, in *Dolphin Cognition and Behavior: A Comparative Approach* (eds. RJ Schusterman, JA Thomas and FG Wood), Laurence Erlbaum Associates, Hillsdale, New Jersey, pp. 167–81.

MM Merzenich, WM Jenkins, P Johnston, C Schreiner, SL Miller and P Tallal (1996) Temporal processing deficits of language-learning impaired children ameliorated by training, *Science*, 271(5245):77–80

GA Miller and PM Gildea (1987) How children learn words, *Sci. Am.*, 257(3):86–91.

T Miyasaka (1995) Bacteriorhodopsin-based artificial photoreceptors, in *Molecular Biology and Biotechnology* (ed. RA Meyers), VCH, New York, pp. 70–3.

Y Miyashita (1995) How the brain creates imagery: projection to primary visual cortex, *Science*, 268(5218):1719–20.

H Moravec (1988) *Mind Children*, Harvard University Press, Cambridge, Massachusetts.

HJ Morowitz (1993) *Beginning of Cellular Life*, Yale University Press, New Haven, Connecticut.

MM Mortland (1984) Deamination of glutamic acid by pyridoxal phosphate-Cu^{2+}-ion-smectite catalysts, *J. Mol. Catal.*, 27:143–55.

BA Moskowitz (1978) The acquisition of language, *Sci. Am.*, 239(5):82–96.

C Nüsslein-Volhard (1996) Gradients that organize embryo development, *Sci. Am.*, 275(2):54–5, 58–61.

KK Obermeier and JJ Barron (1989) Time to get fired up, *Byte*, August:217–24.

J O'Connell (1991) Peirce, Charles Sanders, in *Encyclopedia of Western Philosophy and Philosophers*, 2nd edn (eds. JO Urmson and J Ree), Routledge, London.

MV Olson (1995) A time to sequence, *Science*, 270(5235):394–6.

CS Peirce (1958) in Charles S. Peirce Selected Writings (ed. PP Wiener), Dover, New York.

TJ Pinnavaia and MM Mortland (1986) Aspects of clay catalysis, in *Clay Minerals and the Origin of Life* (eds. AG Cairns-Smith and H Hartman), Cambridge University Press, Cambridge, pp. 131–5.

W Poundstone (1985) *The Recursive Universe*, Contemporary Books, Chicago.

RR Provine (1996) Laughter, *Am. Sci.*, 84(1):38–45.

WE Reichardt and T Poggio (eds.) (1981) *Theoretical Approaches in Neurobiology*, MIT Press, Cambridge, Massachusetts.

C Renfrew (1989) The origins of Indo-European languages, *Sci. Am.*, 260(4):186–94.

I Rosenfield (1988) *The Invention of Memory*, Basic Books, New York.

EL Rossi (1993) The Psychobiology of Mind-Body Healing, WW Norton & Co., New York.

JE Rothman and L Orci (1996) Budding vesicles in living cells, *Sci. Am.*, 274(3):70–5.

D Rumbaugh and S Savage-Rumbaugh (1978) Chimpanzee language research: status and potential, *Behav. Res. Meth. Instrum.*, 10(2):119–31.

ES Savage-Rumbaugh, DM Rumbaugh and S Boysen (1980) Do apes use language?, *Am. Sci.*, 68:49–61.

J Schimenti (1995) Mammalian genome, in *Molecular Biology and Biotechnology* (ed. RA Meyers), VCH, New York, pp. 524–9.

E Schrödinger (1944) (combined reprint, 1967), *What is Life?*, Cambridge University Press, Cambridge.

CE Shannon (1948) A mathematical theory of communication, Bell Syst. Tech. J., 27:379, 623.

CE Shannon and W Weaver (1949, 1964) *The Mathematical Theory of Communication*, University of Illinois Press, Urbana.

JA Shapiro (1988) Bacteria as multicellular organisms, *Sci. Am.*, 258(6):62–9.

GM Shepherd (1990) *The Synaptic Organization of the Brain*, Oxford University Press, Oxford.

V Shevoroskin (1990) How linguists have reconstructed the ancestors of all living languages, *The Sciences* (New York Acad. Sci.), May/June:20–7.

A Sloman (1996), A systems approach to consciousness, *RSA Journal*, 144(5470):40–6.

W Slukin (1965) *Imprinting and Early Learning*, Aldine Press, Chicago, Illinois.

S Sonea (1988) The global organism, *The Sciences* (New York Acad. Sci.), July/August:38–45.

CG Starr (1991) *A History of the Ancient World*, 4th edn, Oxford University Press.

TO Stevens and JP McKinley (1995) Lithoautotrophic microbial ecosystems in deep basal aquifers, *Science*, 270:450–4.

EI Stiefel (1996) Molybdenum bolsters the bioinorganic brigade, *Science*, 272(5268):1599–600.

T Stonier (1983) *The Wealth of Information*, Thames/Methuen, London.

T Stonier (1984) Computer psychology, *Educat. Child Psychol.*, 1(2):16–27.

T Stonier (1986a) Towards a new theory of information, *Telecommunications Policy*, 10(4):278–81.

T Stonier (1986b) What is information?, in *Research and Education in Expert Systems III* (ed. MA Bramer), Cambridge University Press, Cambridge, pp. 217–30.

T Stonier (1988) Machine intelligence and the long-term future of the human species, *AI & Society*, 2:133–9.

T Stonier (1989) Towards a general theory of Information II: information and entropy, *Aslib Proc.*, 41(2):41–55.

T Stonier (1990a) Towards a general theory of information: information and entropy, in *Future Computing Systems*, 2(4): 409–27.

T Stonier (1990b) *Information and the Internal Structure of the Universe*, Springer-Verlag, London.

T Stonier (1992) *Beyond Information: The Natural History of Intelligence*, Springer-Verlag, London.

T Stonier (1996) Information as a basic property of the universe, *BioSystems*, 38(2,3):135–40.

EM Stricker and JG Verbalis (1988) Hormones and behavior: the biology of thirst and sodium appetite, *Am. Sci.*, 76:261–7.

CB Stringer (1988) Homo sapiens, in Encyclopedia of Human Evolution and Prehistory (eds. I Tattersall, E Delson and J Van Couvering), Garland, New York, pp. 267–75.

TT Struhsaker (1967) Auditory communication among vervet monkeys (*Cercopithecus aethiops*), in *Social Communication Among Primates* (ed. SA Altmann), University of Chicago Press, Chicago, Illinois.

L Szilard (1929) Über die Entropieverminderung in einem thermodynamischen System bei Eingriffen intelligenter Wesen, *Z. Physik.*, 53:840–56.

P Tallal, SL Miller, G Bedi, G Byma, X Wang, SS Nagarajan, C Schreiner, WM Jenkins and MM Merzenich (1996), Language comprehension in language-learning impaired children improved with acoustically modified speech, *Science*, 271(5245):81–4.

A Tauber (1994) *The Immune Self*, Cambridge University Press, Cambridge.

G Taubes (1994) Small army of code-breakers conquers a 129-digit giant, *Science*, 264:776–7.

JG Taylor and CLT Mannion (eds.) (1992) *Coupled Oscillating Neurons*, Springer-Verlag, London.

MFJ Taylor, Y Shen and ME Kreitman (1995) A population genetic test of selection at the molecular level, *Science*, 270:1497–9.

P Teilhard de Chardin (1956) *Man's Place in Nature* (Engl. transl. 1966, Fontana Books 1971) William Collins & Sons, London.

P Teilhard de Chardin (1959) *The Future of Man* (Engl. transl. N Denny 1964), William Collins & Sons, London.

WH Thorpe (1961) *Bird-song*, Cambridge University Press, London.

L Tisza (1991) Entropy, in *Encyclopedia of Physics*, 2nd edn (eds. RG Lerner and GL Trigg), VCH, New York, pp. 353–4.

T Toffoli and N Margolus (1987) *Cellular Automata Machines*, MIT Press, Cambridge, Massachusetts.

P Tompkins and C Bird (1973) *The Secret Life of Plants*, Avon Books, New York.

RH Tuttle (1990) Apes of the World, *Am. Sci.*, 78:115–25.

B Vickery and A Vickery (1992) *Information Science*, 2nd edn, Bowker-Saur, London.

H Vöchting (1878) *Über Organbildung im Pflanzenreich*, Max Cohen, Bonn.

F de Waal (1989) *Peacemaking among Primates*, Harvard University Press, Cambridge, Massachusetts.

HG Wells (1915) *Boon, The Mind of the Race, The Wild Asses of the Devil* and *The Last Trump*, T. Fisher Unwin, London.

J Wicken (1987) Entropy and information: suggestions for a common language, *Philos. Sci.*, 54:176–93.

N Wiener (1948) *Cybernetics. Or Control and Communication in the Animal and the Machine*, MIT Press, Cambridge, Massachusetts.

N Wiener (1961) *Cybernetics. Or Control and Communication in the Animal and the Machine*, 2nd edn, MIT Press and John Wiley & Sons, New York.

T Winograd and F Flores (1987) *Understanding Computers and Cognition*, Alex Publ. Corp., Norwood, New Jersey.

JR Wolpaw, JT Schmidt and TM Vaughan (eds.) (1991) *Activity-Driven CNS Changes in Learning and Development*, Ann. New York Acad. Sci. Vol. 627.

HP Yockey (1992) *Information Theory and Molecular Biology*, Cambridge University Press, Cambridge, UK.

HP Yockey (1995) Comments on 'Let there be life; Thermodynamic reflections on biogenesis and evolution' by Avshalom C. Elitzur, *J. Theor. Biol.*, 176:349–55.

Index